English Medieval Church Towers

ENGLISH MEDIEVAL CHURCH TOWERS

THE NORTHERN PROVINCE

W. E. David Ryan

THE BOYDELL PRESS

First published 2018
The Boydell Press, Woodbridge

ISBN 978 1 78327 353 9

The Boydell Press is an imprint of Boydell & Brewer Ltd
PO Box 9, Woodbridge, Suffolk IP12 3DF, UK
and of Boydell & Brewer Inc.
668 Mt Hope Avenue, Rochester, NY 14620–2731, USA
website: www.boydellandbrewer.com

A CIP catalogue record for this book is available
from the British Library

This publication is printed on acid-free paper

Designed and typeset by BBR Design, Sheffield
Printed and bound in the UK by Severn, Gloucester

This book is dedicated to Kathleen for her love and support over more than half a century.

Contents

The Church Towers

Figures

Preface

My interest in medieval church towers was first sparked when I began my career as an architect. Our training included spending time at an art college, and it was here that I first discovered the history of architecture, studying buildings from the Temple of Khonsu at Karnak to Eero Saarinen's Design Dome, by way of St Paul's Cathedral, Blenheim Palace, and nineteenth-century cotton mills! However, my attention was always most held by buildings from the Middle Ages – and particularly religious buildings.

After retirement (partly prompted by the enforced introduction of computer-aided design), I was able to spend more time investigating these wonderful buildings, and towers in particular, with the help both of Pevsner's *Buildings of England* and the navigational skills of my wife, Kathleen. The watercolours in this book are the fruits of my explorations in the ecclesiastical province of York (comprising the dioceses of Blackburn, Carlisle, Chester, Durham, Leeds, Liverpool, Manchester, Newcastle, Sheffield, Southwell and Nottingham, and York); they cover an area stretching from Ancroft (Northumberland) in the north to Stanford on Soar (Nottinghamshire) in the south, and from Dearham (Cumbria) in the west to Easington (Yorkshire) in the east.

The reproductions of my paintings are accompanied by text extracted and adapted from the National Heritage List for England and I am grateful to Historic England for kindly granting permission to reproduce this material.

The National Heritage List text entries contained in this material were obtained on 10 November 2016 and are © Historic England 2016. The most publicly available up-to-date National Heritage List text entries can be obtained from http://www.historicengland.org.uk/listing/the-list/.

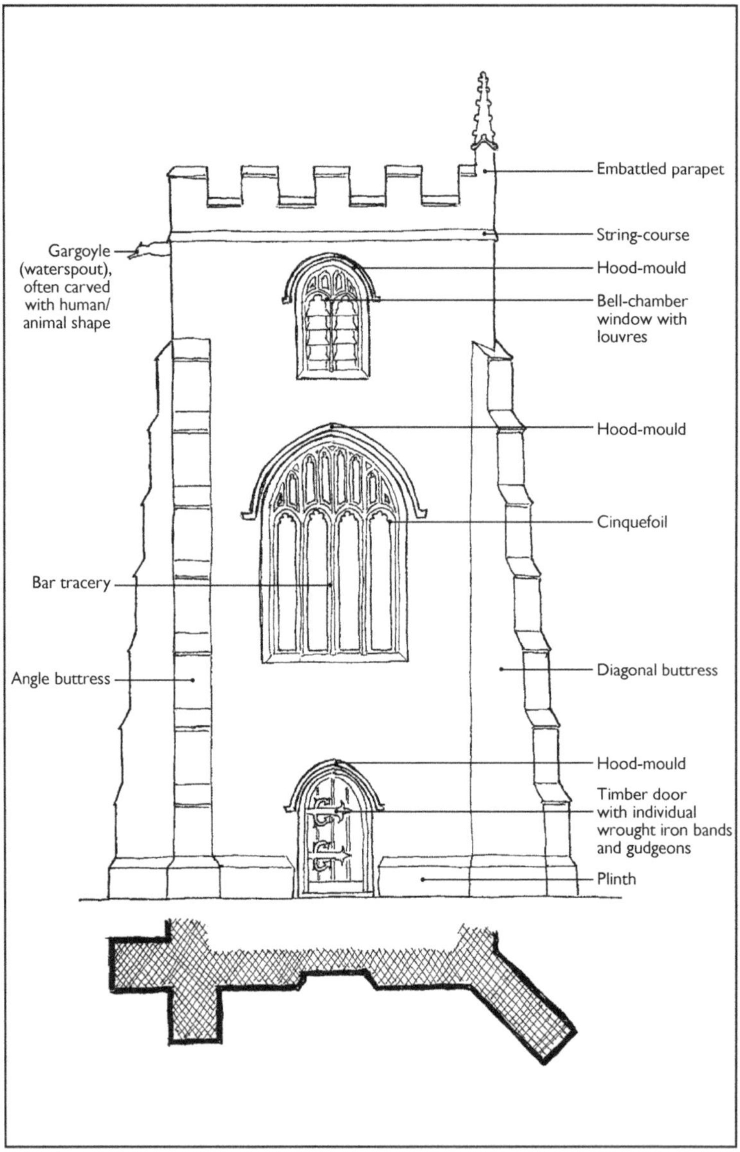

Figure 1. West elevation of a typical medieval church tower

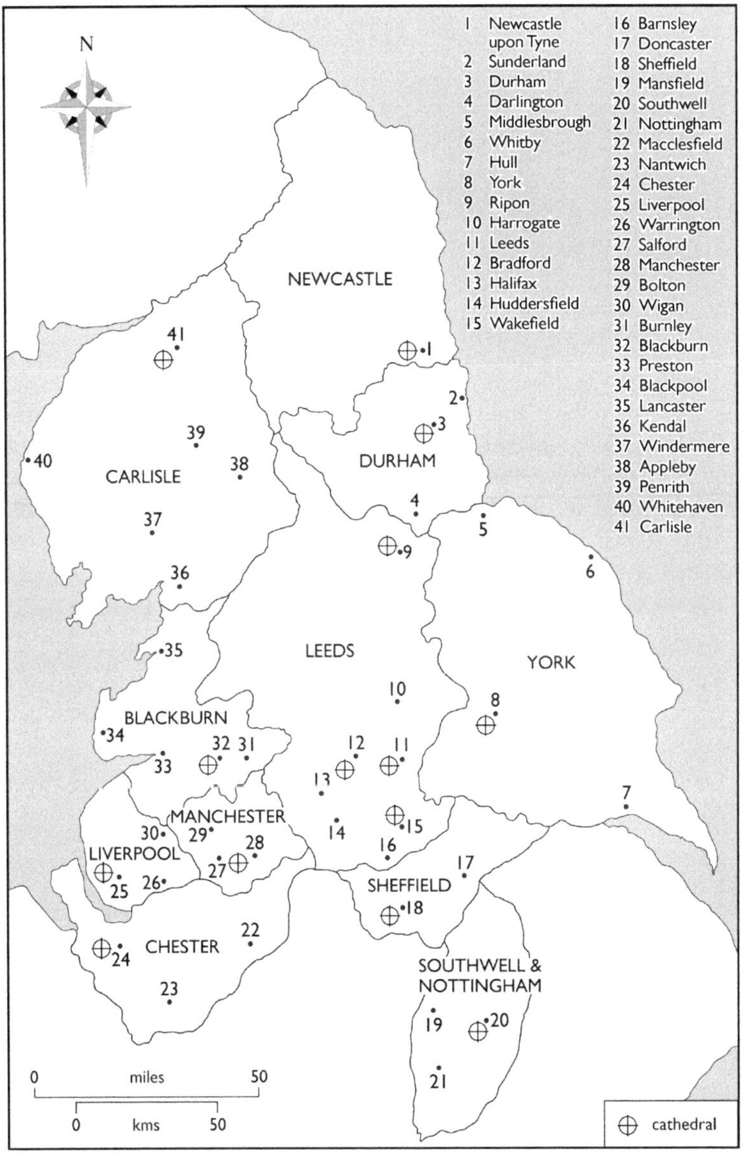

Figure 2. Map of the Northern Province

Acknowledgements

I would like to place on record my sincere thanks to the following persons whose help and support was of great value and is much appreciated: Kathleen, my wife, for being ever-present with encouragement and map-reading abilities; Reverend Canon Roy McCullough MBE, for his continued and informative enthusiasm; Margaret Ryan along with Philip, Richard and Laurence Howard for their IT assistance; and at Boydell & Brewer, Caroline Palmer, for her imaginative foresight, and Robert Kinsey, for his literary skills allied to his attention to detail.

W. E. David Ryan

THE CHURCH TOWERS

Diocese of

Blackburn

1 Bolton-by-Bowland St Peter & St Paul

LOCATION: Gisburn Road, Bolton-by-Bowland

POSTCODE: BB7 4NP

OS GRID REF: SD786494

DATE: Mid-C15 (nave and tower said to be completed 1466/67)

DESCRIPTION: Sandstone ashlar west tower, diagonal buttresses with ornamented offsets, embattled parapet with corner pinnacles, and gargoyles. Bell openings in 2 stages. Upper ones have Tudor-arched heads and hoods with stops. The 2 cusped lights of each opening are separated by a splayed mullion which projects forward and continues upwards across the face of the parapet. Above each opening are decorative motifs including pierced quatrefoils. The lower bell-openings are of 2 cusped lights with panel tracery above, under pointed heads with hoods having head stops. The west window is of 4 cusped lights with panel tracery, under a Tudor-arched head with hood and decorative stops. The door has a pointed Tudor arch with hood and head stops, and with jambs of 2 hollow-chamfered orders.

GRADE: I

HERITAGE LIST NUMBER: 1362287

2 Bolton-le-Sands Holy Trinity

LOCATION: Main Road, Bolton-le-Sands

POSTCODE: LA5 8DP

OS GRID REF: SD483677

DATE: Late C15

DESCRIPTION: Sandstone ashlar west tower of 3 stages with diagonal buttresses and battlemented parapet. Bell openings of 3 pointed lights under a 4-centred head with hood-mould. Over the west door is a similar window. The west doorway, of 2 hollow-chamfered orders, has a 2-centred head and hood.

GRADE: II*

HERITAGE LIST NUMBER: 1071944

3 Brindle St James

LOCATION: Sandy Lane, Brindle

POSTCODE: PR6 8NH

OS GRID REF: SD599242

DATE: c.1500

DESCRIPTION: Late medieval Perpendicular sandstone west tower. 4 stages with chamfered plinth, diagonal buttresses and south-east stair-turret. Tudor-arched west doorway with hollowed surround, above this a restored 3-light window under a hood-mould with shield stops, a clock face, and 2-light louvred belfry windows with Perpendicular tracery and hood-moulds; on north and south sides chamfered loop-lights, and gargoyles below a drip-course carried round the whole; embattled parapet with crocketed corner pinnacles.

GRADE: II

HERITAGE LIST NUMBER: 1072542

4 Broughton St John Baptist

LOCATION: Church Lane, Broughton

POSTCODE: PR3 5LN

OS GRID REF: SD528343

DATE: 1533

DESCRIPTION: Sandstone, 3-stage embattled tower with diagonal buttresses (dying in the height, with offsets) and full-height south-west stair-turret, has a moulded arched west door and above this a 3-light Perpendicular window with tracery, a 3-light belfry louvre in each side; loop-lights to the stair-turret, and a parapet string-course dated 1533 on the south side, with small gargoyles near the corners.

GRADE: II*

HERITAGE LIST NUMBER: 1164208

5 Burnley St Peter

LOCATION: Church Street, Burnley

POSTCODE: BB11 2DL

OS GRID REF: SD842329

DATE: C15 (raised 1802)

DESCRIPTION: Sandstone west tower, of 4 unequal stages with a massive moulded plinth, angle buttresses to half-height, a drip-band over the third stage, a set-back fourth stage with a corbel-table, plain cornice and embattled parapet with intermediate and corner pinnacles, and an elaborate weathervane. Has a 2-centred arched west doorway with hollow-moulded surround, hood-mould with figured stops (worn) and double doors with scrolled strap hinges; a restored 3-light west window with cusped lights, Perpendicular tracery and a hood-mould; a similar but smaller 2-light window to the third stage; clock faces in the top stage with small 2-centred arched 1-light belfry windows above; and at the south-east corner a narrow rectangular vice (or stair-turret) which has small chamfered loop-lights.

GRADE: II*

HERITAGE LIST NUMBER: 1022640

6 Caton St Paul

LOCATION: Brookhouse Road, Caton

POSTCODE: LA2 9JG

OS GRID REF: SD542646

DATE: Probably C16

DESCRIPTION: Snecked sandstone rubble west tower of 3 stages with diagonal buttresses and an embattled parapet. The bell-openings are of 3 round-headed lights under a flat head with hood. The west door has a pointed head with 2 hollow-chamfered orders, and a hood. Above is a window of 3 cusped lights with Perpendicular tracery under a segmental head.

GRADE: II*

HERITAGE LIST NUMBER: 1163957

7 Chorley St Laurence

LOCATION: Union Street, Chorley

POSTCODE: PR7 1EA

OS GRID REF: SD582177

DATE: C15

DESCRIPTION: Battlemented west tower of 3 stages has diagonal west buttresses each with 3 carved boar heads at first stage and cusped niche at second stage; between these buttresses on first-floor drip-mould with various figured bosses in the casement (3 faces, 2 animal heads, flowers), a west doorway (probably C20) and a C19 4-light west window with curvilinear tracery and a rose in the head; third stage has an arched 2-light belfry louvre with hood-mould in each side, above which the north and south sides have prominent gargoyles; slightly projecting stair-turret on south side.

GRADE: II*

HERITAGE LIST NUMBER: 1072631

8 Church Kirk
St James

LOCATION: St James Road, Church Kirk

POSTCODE: BB5 4JP

OS GRID REF: SD741291

DATE: C15

DESCRIPTION: Embattled west tower of uncoursed roughly hewn blocks. Has diagonal (west) and angle (east) buttresses, drip-courses above and below the belfry, low rectangular west door with moulded jambs and lintel, and a hood-mould, a 3-light Perpendicular window with a gabled hood-moulds, similar but smaller openings to 3 sides of belfry (clock on fourth side), and various small 1-light openings.

GRADE: II*

HERITAGE LIST NUMBER: 1072728

9 Churchtown
St Helen

LOCATION: 17 Church Street, Churchtown

POSTCODE: PR3 0HT

OS GRID REF: SD413427

DATE: C15

DESCRIPTION: The tower has diagonal buttresses, a stair-turret with stone spirelet, and an embattled parapet. The bell-openings are of 2 trefoiled lights under a flat head with hood. The west window is of 3 trefoiled lights with pointed head and Perpendicular tracery. The west door has a pointed head.

GRADE: I

HERITAGE LIST NUMBER: 1072874

10 Cockerham St Michael

LOCATION: Marsh Lane, Cockerham

POSTCODE: LA2 0EQ

OS GRID REF: SD462517

DATE: C16 ("1589" said to be inscribed on beam from tower).

DESCRIPTION: 3-stage west tower with diagonal buttresses, a stair-turret on the south side, and an embattled parapet. The bell-openings are each of 3 pointed lights under a Tudor-arched head with hood. The west window is similar. The west doorway has a round arch, hollow-chamfered in 2 orders.

GRADE: II*

HERITAGE LIST NUMBER: 1317937

11 Colne St Bartholomew

LOCATION: Church Street, Colne

POSTCODE: BB8 0BU

OS GRID REF: SD888402

DATE: Early C16

DESCRIPTION: West tower with moulded plinth and offset angle buttresses to western corners. Pointed-arch west door and 3-light window above, flanked by 4 shields. 3-light pointed transomed belfry windows. Square south-east stair-turret and oversailing embattled parapet.

GRADE: I

HERITAGE LIST NUMBER: 1073414

12 Croston
St Michael & All Angels

LOCATION: Church Street, Croston

POSTCODE: PR26 9HA

OS GRID REF: SD490184

DATE: C15

DESCRIPTION: A massive square 4-stage tower, mostly embraced by nave, with angle buttresses and embattled parapet with corner and intermediate pinnacles, has deep moulded pointed-arch west door with hood-mould terminating in carved heads, above this an arched 4-light window with moulded surround (probably C16 but containing later tracery), 3-light belfry louvres with intersecting tracery and hood-moulds; clock faces on the north and south sides; and loop-lights to an internal vice in the south-west corner.

GRADE: II*

HERITAGE LIST NUMBER: 1163631

13 Downham
St Leonard

LOCATION: Chatburn Road, Downham

POSTCODE: BB7 4BN

OS GRID REF: SD784443

DATE. Late C15

DESCRIPTION: The tower has diagonal buttresses with offsets and an oversailing embattled parapet with corner pinnacles and with 2 gargoyles on both the north and south sides. The bell-openings are of 2 cinquefoiled lights with outer casement moulding, under a pointed head with hood. The west window has a pointed Tudor-arched head with hood, an outer casement moulding, and 3 cinquefoiled lights. The west door has a pointed head, a casement moulding, and a hood.

GRADE: II*

HERITAGE LIST NUMBER: 1164023

14 Eccleston St Mary

LOCATION: Towngate, Eccleston

POSTCODE: PR7 5QL

OS GRID REF: SD516178

DATE: C14 with C15 bell stage and C18 parapet

DESCRIPTION: A square red sandstone tower of 3 stages, mostly C14 but with C15 bell stage and C18 parapet. The tower has diagonal west buttresses up to the second stage, which terminates in a chamfered coping, the bell stage set back and finished with a parapet of ashlar; at ground floor in place of a west door is a modern window of 2 cusped lights, at second stage a clock face and on the south side a little segmental-arched window to the ringing chamber; the belfry has louvred windows of 2 cusped lights and multifoil head, in moulded surround, with hood-moulds; straight parapet has panelled corner posts bearing carved scallops each with a small weathervane.

GRADE: II*

HERITAGE LIST NUMBER: 1362129

15 Gisburn St Mary

LOCATION: Main Street, Gisburn

POSTCODE: BB7 4HQ

OS GRID REF: SD829489

DATE: C14 with Norman base

DESCRIPTION: Sandstone rubble west tower. The tower has no buttresses and has an embattled parapet with corner pinnacles. The bell-openings are of 2 trefoiled lights under a pointed head. The west opening is possibly an early C19 copy. The west door and west window appear to be C19. The door is chamfered with a shouldered flat head, and the window of 3 lights.

GRADE: II*

HERITAGE LIST NUMBER: 1164472

16 Goosnargh St Mary the Virgin

LOCATION: Goosnargh Lane, Goosnargh

POSTCODE: PR3 2BJ

OS GRID REF: SD559369

DATE: C15

DESCRIPTION: Sandstone rubble 3-stage tower, with full-height south-east stair-turret and diagonal 6-stage west buttresses, has on west side a round-headed chamfered doorway with hood-mould, a 3-light window with Perpendicular tracery; on west and south sides diamond-shaped clock faces; and on all sides 3-light louvred belfry windows with Perpendicular tracery, and an embattled parapet.

GRADE: II*

HERITAGE LIST NUMBER: 1361634

17 Great Harwood St Bartholomew

LOCATION: Church Lane, Great Harwood

POSTCODE: BB6 7WN

OS GRID REF: SD734327

DATE: C15

DESCRIPTION: Embattled Perpendicular tower of 4 short stages has diagonal (west) and angle (east) buttresses, full-height rectangular stair-turret at south-east corner, west window with hood-mould, a niche above this, 2-light belfry openings with hood-moulds (clock faces inserted in those on west and south sides), and various slit-shaped openings.

GRADE: II*

HERITAGE LIST NUMBER: 1362006

18 Halton St Wilfrid

LOCATION: Church Brow, Halton

POSTCODE: LA2 6LP

OS GRID REF: SD498647

DATE: 1597

DESCRIPTION: C19 church with surviving west tower said to be 1597. Tower of 3 stages, battlemented with finials. All tower windows and openings have hood-moulds and 4-centred heads to the lights. Each face has a 2-light bell-opening. The south wall has a 1-light window on the first and second floors. The west wall has a C19 doorway with a 2-light window above and a 1-light window above that. The north wall has a stair-turret.

GRADE: II

HERITAGE LIST NUMBER: 1164098

19 Hornby St Margaret

LOCATION: Main Street, Hornby

POSTCODE: LA2 8LZ

OS GRID REF: SD585685

DATE: 1514

DESCRIPTION: Sandstone west tower built in 1514 by Sir Edward Stanley Lord Monteagle. The tower is octagonal in plan and of 3 stages, the 2 upper stages being set diagonally to the base. The parapet is embattled with pinnacles and the bell-openings each have a mullion and a transom and have Tudor-arched heads with hoods. Above and below them is a string with corner gargoyles. The middle stage has a plaque carved with the Monteagle arms. The west window is of 3 round-headed lights under a pointed head with Perpendicular tracery. Above is a panel inscribed: "E. Stanley : miles : dnu : Montegle . me fieri fecit" (E. Stanley, knight, Lord Monteagle had me made). The west doorway is chamfered in 2 orders and has a pointed head with hood.

GRADE: I

HERITAGE LIST NUMBER: 1071657

20 Leyland St Andrew

LOCATION: Church Road, Leyland

POSTCODE: PR25 3FJ

OS GRID REF: SD541216

DATE: C15

DESCRIPTION: Embattled west tower of 4 stages, with diagonal buttresses to the west and angle buttresses to the east, moulded plinth; west door in low Perpendicular arch with hood-mould, west window with C19 tracery, above this is a small niche with moulded jambs and head, and above this again a small round-headed opening (similar niches and openings on north and south sides, and on the south 6 vertically arranged illumination slits for staircase) clock faces on 3 sides; belfry windows on all sides, each of 3 lights under low pointed arch; drip-mould, with gargoyles on 3 sides, has on west side carvings of a woodcock and 4-petalled flowers (Seth Woodcock was vicar of Leyland, 1488–1516).

GRADE: II*

HERITAGE LIST NUMBER: 1073036

21 Melling
St Wilfrid

LOCATION: Main Street, Melling

POSTCODE: LA6 2RR

OS GRID REF: SD598711

DATE: Late C15

DESCRIPTION: Sandstone rubble west tower with diagonal buttresses and an embattled parapet. The bell-openings are of 3 pointed lights under a Tudor-arched head with hood. The west window is similar. The west doorway has a pointed head with hood and is hollow-chamfered in 2 orders.

GRADE: I

HERITAGE LIST NUMBER: 1165114

22 Mitton
All Hallows

LOCATION: Clitheroe Road, Mitton

POSTCODE: BB7 9PL

OS GRID REF: SD715389

DATE: Early C15

DESCRIPTION: Sandstone rubble 3-stage west tower with diagonal buttresses, and an embattled parapet. The bell-openings have pointed heads with hoods, and 2 cinquefoiled lights with panel tracery. The west window has 4 cinquefoiled lights, panel tracery, and a hood. The west doorway is wide and moulded with a pointed head and hood.

GRADE: I

HERITAGE LIST NUMBER: 1163432

23 Newchurch in Pendle St Mary

LOCATION: Cross Lane, Newchurch in Pendle

POSTCODE: BB12 9JP

OS GRID REF: SD823394

DATE: 1653

DESCRIPTION: Tower of 2 stages separated by string. Large squared quoins, 2 of which are inscribed "ID 1653" and "IH CC BS NR". At ground level, above moulded plinth, is small 2-light chamfered mullioned window: the belfry windows, which now have flat-faced mullions, may once have been similar. Embattled, parapet probably C19.

GRADE: II*

HERITAGE LIST NUMBER: 1361745

24 Over Kellet St Cuthbert

LOCATION: Nether Kellet Road, Over Kellet

POSTCODE: LA6 1DN

OS GRID REF: SD522696

DATE: Early C16

DESCRIPTION: The tower has diagonal buttresses and a battlemented parapet. In the west wall is a blocked doorway with double hollow chamfer, 2-centred head and hood-mould. Above is a 3-light C16 window with hood-mould. Bell openings are all C16, of 3 lights.

GRADE: II*

HERITAGE LIST NUMBER: 1071877

25 Penwortham St Mary

LOCATION: Church Avenue, Penwortham

POSTCODE: PR1 0AH

OS GRID REF: SD524290

DATE: C15

DESCRIPTION: Battlemented west tower with diagonal buttresses; west doorway with moulded arch and hood-mould; moulded dripstone band (west front only) from which rises an arched 3-light first-floor window with Perpendicular tracery and hood-mould; above this is an ogee-headed niche with crocketed pinnacles, and at top level on all sides arched and traceried 2-light belfry louvres with hood-moulds; battlemented coping with pinnacles.

GRADE: II*

HERITAGE LIST NUMBER: 1073058

26 Ribchester St Wilfrid

LOCATION: Riverside, Ribchester

POSTCODE: PR3 3XS

OS GRID REF: SD649350

DATE: Late C15

DESCRIPTION: 3-stage sandstone rubble tower, said to be late C15, with angle buttresses and battlements. The bell-openings are of 3 lights with hoods and with panel tracery to their heads. The west wall has a similar window above the doorway, which has a 2-centred head and a moulding similar to that of the south porch door.

GRADE: I

HERITAGE LIST NUMBER: 1147451

27 St Michael's on Wyre
St Michael

LOCATION: Blackpool Road, St Michael's on Wyre

POSTCODE: PR3 0UQ

OS GRID REF: SD462410

DATE: 1549

DESCRIPTION: 3-stage sandstone ashlar tower, said to be of 1549, is of trapezoidal plan and has angle buttresses. The upper stage is set back and has an embattled parapet with corner pinnacles. The date 1611, carved on a merlon together with the Butler arms, probably refers to a raising of the tower. The bell-openings are each of 2 chamfered cusped lights. The west window is mullioned and transomed, of 6 lights under a Tudor-arched head with hood. The west doorway is hollow-chamfered in 2 orders, with Tudor-arched head and hood.

GRADE: I

HERITAGE LIST NUMBER: 1281178

28 Tunstall
St John Baptist

LOCATION: Church Lane, Tunstall

POSTCODE: LA6 2RQ

OS GRID REF: SD614739

DATE: c.1415

DESCRIPTION: Sandstone rubble west tower with diagonal buttresses and an embattled parapet. The bell-openings are of 2 trefoiled lights with flat heads and hoods. Above each opening, above a weathered set-back, is a carved tablet of an angel holding a shield. The west window is of 3 cusped lights under a pointed head with Perpendicular tracery. The west door has a hollow-chamfered surround with pointed head and hood.

GRADE: I

HERITAGE LIST NUMBER: 1071642

29 Waddington St Helen

LOCATION: Edisford Road, Waddington

POSTCODE: BB7 3HS

OS GRID REF: SD728438

DATE: c.1500

DESCRIPTION: 3-stage sandstone tower with angle buttresses, an embattled parapet with string-course, gargoyles and corner pinnacles, and bell-openings with pointed heads and hoods and 2 cusped lights with panel tracery above. The west wall has a window of 3 cusped lights with panel tracery. The west door is deeply chamfered with pointed head and hood. The north-west buttress has a worn carved crucifix and arms of the Tempest family.

GRADE: II*

HERITAGE LIST NUMBER: 1163679

30 Walton-le-Dale St Leonard

LOCATION: Church Brow, Walton-le-Dale

POSTCODE: PR5 4HJ

OS GRID REF: SD561281

DATE: C16

DESCRIPTION: 4-stage tower with angle buttresses, west door with moulded arch and above this a 3-light arched window with Perpendicular tracery and hood-mould, attached octagonal clock face on west side, arched 3-light belfry louvres with Perpendicular tracery and hood-moulds on all sides; drip-mould and embattled parapet.

GRADE: II*

HERITAGE LIST NUMBER: 1074102

31 Warton
St Oswald

LOCATION: Main Street, Warton

POSTCODE: LA5 9PH

OS GRID REF: SD498723

DATE: Late C15

DESCRIPTION: Pebble-dashed west tower with sandstone dressings. The tower has diagonal buttresses, a late C19 battlemented parapet with pinnacles and 3-light bell-openings of similar date. The west wall has a doorway of 2 chamfered orders with hood-mould, now partly blocked.

GRADE: II

HERITAGE LIST NUMBER: 1308887

32 Whalley
St Mary & All Saints

LOCATION: Church Lane, Whalley

POSTCODE: BB7 9SY

OS GRID REF: SD732361

DATE: C15

DESCRIPTION: Sandstone rubble west tower with angle buttresses and an embattled parapet. The bell-openings are of 2 trefoiled lights with tracery and hoods. The west window is of 3 trefoiled lights under a pointed head with Perpendicular tracery. Below is a moulded doorway with pointed arch.

GRADE: I

HERITAGE LIST NUMBER: 1164684

33 Whittington St Michael the Archangel

LOCATION: Church Street, Whittington

POSTCODE: LA6 2NU

OS GRID REF: SD600763

DATE: Probably early C16

DESCRIPTION: Sandstone rubble west tower with diagonal buttresses with offsets, and an embattled parapet. The bell-openings each have 3 round-headed lights under a flat head. Below on the west wall is a niche with statue. The west window is of 3 round-headed lights under a flat head with hood. The west doorway is pointed and hollow-chamfered in 2 orders.

GRADE: II*

HERITAGE LIST NUMBER: 1071615

Diocese of
Carlisle

34 Aldingham St Cuthbert

LOCATION: Coast Road, Aldingham

POSTCODE: LA12 9RT

OS GRID REF: SD 283710

DATE: C14

DESCRIPTION: Tower with diagonal buttresses; west entrance is chamfered, hood has 1 stop, 3-light window above has Perpendicular tracery; 2-light straight-headed bell-openings, but that to north pointed. Coped embattled parapet and angle pinnacles.

GRADE: II*

HERITAGE LIST NUMBER: 1335944

35 Appleby St Lawrence

LOCATION: Boroughgate, Appleby

POSTCODE: CA16 6QN

OS GRID REF: NY683205

DATE: Lower stage c.1150, with C14 and C15 additions

DESCRIPTION: Castellated west tower with clock of 1699.

GRADE: I

HERITAGE LIST NUMBER: 1312067

36 Arthuret St Michael & All Angels

LOCATION: Arthuret Road, Longtown

POSTCODE: CA6 5SH

OS GRID REF: NY379677

DATE: Completed c.1700. Restored 1868.

DESCRIPTION: 3-storey red sandstone tower. Plank door in pointed arch, 2-light mullioned openings with louvred vents; small, chamfered-surround stair windows to left; stepped angle buttresses, string-courses, battlemented parapet and corner pinnacles; weathervane with cut lettering.

GRADE: II*

HERITAGE LIST NUMBER: 1087523

37 Beetham St Michael

LOCATION: Church Street, Beetham

POSTCODE: LA7 7AL

OS GRID REF: SD496795

DATE. Probably C12. Top stage added in C16.

DESCRIPTION: 2-stage limestone and sandstone rubble battlemented west tower with crocketed corner pinnacles. C14 doorway with 2-centred arch in west wall and window above with 2 trefoiled ogee lights in 2-centred head with moulded label; loop-lights to all but east side; bell-chamber, slightly corbelled out, has 3-light C16 windows with elliptical-headed lights under moulded labels.

GRADE: I

HERITAGE LIST NUMBER: 1137740

38 Brough St Michael

LOCATION: Church Brough, Brough

POSTCODE: CA17 4EJ

OS GRID REF: NY793139

DATE: 1513–25

DESCRIPTION: 3-stage west tower built by the Blenkinsops of Helbeck, 1513–25.

GRADE: II*

HERITAGE LIST NUMBER: 1144902

39 Burgh by Sands St Michael

LOCATION: Centre of village, Burgh by Sands

POSTCODE: CA5 6AW

OS GRID REF: NY328591

DATE: c.1360

DESCRIPTION: 3-stage west tower of red calciferous sandstone. The tower has extremely thick walls on a chamfered plinth with clasping buttresses; vaulted lower chamber has newel staircase in south-west angle, lit by arrow slits; west wall has internal steps in thickness of wall, to arrow slit; loop hole near former north entrance in aisle. South-east buttress has inscription "I.S. 1560" (?). First-floor trefoil-headed lancets in each wall, that in east wall looks into nave; round-arched bell-openings above, square-headed in east wall. Battlemented parapet with projecting lead waterspouts. Ordnance Survey benchmark on north-west buttress. East entrance from nave has iron yet and draw-bar tunnel. Medieval bells.

GRADE: I

HERITAGE LIST NUMBER: 1367134

40 Burton-in-Kendal St James

LOCATION: A6070, north of Burton-in-Kendal

POSTCODE: LA6 1NT

OS GRID REF: SD530769

DATE: C14 with C12 base

DESCRIPTION: 3-stage tower with later embattled parapets, C14 west window with 2 trefoiled ogee lights, square-headed window in north wall and early C16 2-light windows in each wall of bell-chamber.

GRADE: I

HERITAGE LIST NUMBER: 1335703

41 Caldbeck St Kentigern

LOCATION: Centre of village, Caldbeck

POSTCODE: CA7 8DP

OS GRID REF: NY326399

DATE: C12 with C18 upper storey

DESCRIPTION: 3-storey square limestone west tower. Tower is medieval up to string-course with small round-arched window; upper storey is dated by internal inscription "THIS STEEPLE WAS BUILDED IN THE YEAR 1727", with louvred round-arched openings under battlemented parapet.

GRADE: I

HERITAGE LIST NUMBER: 1327205

42 Crosthwaite St Kentigern

LOCATION: Church Lane, Crosthwaite

POSTCODE: CA12 5RA

OS GRID REF: NY257243

DATE: Early C16

DESCRIPTION: 3-stage west tower with diagonal buttresses and embattled parapet incorporating a higher south-west turret. The west doorway is under a C19 shouldered lintel and above it the 4-light window has C19 Perpendicular tracery in a C16 opening. Square-headed belfry openings have triple round-headed lights with sunk spandrels (replaced in wood to the west face), and a big south clock face is on a slate panel.

GRADE: II*

HERITAGE LIST NUMBER: 1327118

43 Dearham St Mungo

LOCATION: Church Street, Dearham

POSTCODE: CA15 7HX

OS GRID REF: NY073364

DATE: c.1300

DESCRIPTION: Square west 3-stage tower that may have been fortified. The tower has a 2-light west window and 2-light bell-openings, but all other windows are small.

GRADE: I

HERITAGE LIST NUMBER: 1144506

44 Grasmere St Oswald

LOCATION: Centre of village, Grasmere

POSTCODE: LA22 9SN

OS GRID REF: NY337074

DATE: C14

DESCRIPTION: Sturdy castellated west tower with a batter and narrow cusped lancet window.

GRADE: I

HERITAGE LIST NUMBER: 1245157

45 Great Salkeld St Cuthbert

LOCATION: Church Lane, Great Salkeld

POSTCODE: CA11 9NA

OS GRID REF: NY552368

DATE: c.1380

DESCRIPTION: 3-storey square west tower of large red sandstone blocks with chamfered plinths. The walls of the tower are extremely thick. Small chamfered loops on each face at various levels, the upper west one with cusped head. Clock face on south wall and projecting angle stair-turret, all under battlemented parapet with angle spout.

GRADE: II*

HERITAGE LIST NUMBER: 1100260

46 Hawkshead
St Michael & All Angels

LOCATION: Main Street, Hawkshead

POSTCODE: LA22 0NT

OS GRID REF: SD352980

DATE: Late C15 or early C16

DESCRIPTION: West tower of stone rubble. The tower has a west pointed entrance, 2-light single-chamfered mullioned window above, and straight-headed louvred bell-opening; top cornice and embattled parapet with angle pinnacles rebuilt 1875. To south a small window with clock face above, below bell-opening.

GRADE: I

HERITAGE LIST NUMBER: 1087259

47 Kendal
Holy Trinity

LOCATION: Kirkland, south end of town

POSTCODE: LA9 5AQ

OS GRID REF: SD516922

DATE: C15

DESCRIPTION: Comprehensive C19 restorations in Perpendicular style. Earlier masonry coursed, squared, rubble; later work snecked rubble. Chamfered plinths and stepped buttresses. Graduated slate roofs with gargoyles and pinnacles.

GRADE: I

HERITAGE LIST NUMBER: 1319009

48 Kirkby Lonsdale St Mary

LOCATION: Queen's Square, Kirkby Lonsdale

POSTCODE: LA6 2AU

OS GRID REF: SD611788

DATE: C12–C16 with alterations C18 and C19

DESCRIPTION: West tower of 4 storeys. Tower (partly C18) and walls entirely embattled in C19. Elaborate Romanesque doorway of 4 orders.

GRADE: I

HERITAGE LIST NUMBER: 1145774

49 Kirkby Stephen St Stephen

LOCATION: Market Square, Kirkby Stephen

POSTCODE: CA17 4QX

OS GRID REF: NY775088

DATE: Early C16

DESCRIPTION: 3-storey west tower. West doorway has continuous chamfered jambs and pointed head in square frame incorporating coats of arms; 3-light window above. Each face has paired bell-openings of 2 lights under ogee gables. Embattled parapet with pinnacles.

GRADE: II*

HERITAGE LIST NUMBER: 1136925

50 Kirkby Thore St Michael

LOCATION: Priest Lane, Kirkby Thore

POSTCODE: CA10 1UR

OS GRID REF: NY638259

DATE: C12 with upper works rebuilt C16

DESCRIPTION: 3-stage tower built in C12 with upper works rebuilt in C16; embattled parapet has C17 gabled bell-cote on east side. Main belfry openings are paired, with trefoil heads.

GRADE: II*

HERITAGE LIST NUMBER: 1226012

51 Long Marton St Margaret & St James

LOCATION: 0.5 mile south of Long Marton

POSTCODE: CA16 6JP

OS GRID REF: NY666240

DATE: Early C12

DESCRIPTION: 3-stage tower added probably early C12. Paired belfry opening on the north side is original, others are C19 reconstructions.

GRADE: I

HERITAGE LIST NUMBER: 1075165

52 Newton Arlosh
St John the Evangelist

LOCATION: B5307, Newton Arlosh

POSTCODE: CA7 5ET

OS GRID REF: NY198553

DATE: C14

DESCRIPTION: Square fortified west tower with extremely thick walls. Licence to crenellate 11 April 1304. Tower has original and restored arrow-slit windows. 2-light Tudor north window has C19 restoration. Upper part of tower is of different stonework and has been restored with battlemented parapet and turret.

GRADE: II*

HERITAGE LIST NUMBER: 1212611

53 Orton
All Saints

LOCATION: B6260, north edge of Orton

POSTCODE: CA10 3RQ

OS GRID REF: NY622084

DATE: Early C16

DESCRIPTION: 3-stage castellated west tower of stone blocks on chamfered plinth with stepped buttresses.

GRADE: II*

HERITAGE LIST NUMBER: 1145428

54 Penrith St Andrew

LOCATION: St Andrew's Place, Penrith

POSTCODE: CA11 7XX

OS GRID REF: NY516301

DATE: C12 and C13

DESCRIPTION: Red sandstone rubble west tower of C12 and C13. The west doorway and rest of church was rebuilt in 1720.

GRADE: I

HERITAGE LIST NUMBER: 1145048

55 Scaleby All Saints

LOCATION: Centre of village, Scaleby

POSTCODE: CA6 4LJ

OS GRID REF: NY447632

DATE: Early C14

DESCRIPTION: 3-storey fortified square west tower. Walls 1.3 metres thick, with slit windows and chamfered plinth. Rebuilt above string-course before 1790, with battlemented parapet of 1827 and corner pinnacles of 1828: square bell-chamber openings, existed before 1790.

GRADE: II

HERITAGE LIST NUMBER: 1335573

56 Sedbergh
St Andrew

LOCATION: Main Street, Sedbergh

POSTCODE: LA10 5BN

OS GRID REF: SD657921

DATE: Mostly c.1500

DESCRIPTION: The square 3-stage west tower has west buttresses to half height, with 4 offsets, belfry stage slightly corbelled out and embattled parapet with corner pinnacles; a restored 2-centred arched 3-light west window with cusped lights, a transom and a hood-mould; the hood-mould of a former window at a high level of the second stage (and a very small opening below this); and a square-headed belfry window of 3 round-headed lights with hollow spandrels, with stone louvres and a hood-mould.

GRADE: I

HERITAGE LIST NUMBER: 1384191

57 Urswick
St Mary the Virgin & St Michael

LOCATION: Church Road, Great Urswick

POSTCODE: LA12 0TA

OS GRID REF: SD268742

DATE: C13 with Perpendicular upper stage

DESCRIPTION: Stone rubble west tower. West entrance with 3-light, Perpendicular window above and C15 pietà in niche; 3-light straight-headed louvred bell-openings; coped embattled parapet.

GRADE: I

HERITAGE LIST NUMBER: 1086801

Diocese of
Chester

58 Audlem St James the Great

LOCATION: Stafford Street, Audlem

POSTCODE: CW3 0AP

OS GRID REF: SJ660436

DATE: Perpendicular incorporating late C13 and early C14 remnants

DESCRIPTION: North-western tower. Western face has angle buttresses with offsets. 2-light lower window with trefoil-headed lights and a quatrefoil to the apex with a hood-mould. Above this are set 2 quatrefoil lights and above them a C19 clock face with an iron outer ring. The stone centre bears an inscription, "J Hervey/J Gouldborne S." The girth of the tower diminishes above this via an offset and the 2-light belfry opening has a casement moulding to the surround and 2 louvred lights with trefoil heads and a quatrefoil to the apex with a hood-mould above. There are gargoyles to the angles and a battlemented parapet with crocketed pinnacles at the corners. The north face has no windows to the lower level, 2 lancet windows at the level of the clock face and is similar to the western face at belfry level and above. The eastern and southern faces adjoin the nave and project only at belfry level and above where they are similar to the western and northern faces.

GRADE: I

HERITAGE LIST NUMBER: 1136872

59 Backford
St Oswald

LOCATION: Church Lane, adjoining Backford Hall

POSTCODE: CH2 4BE

OS GRID REF: SJ398716

DATE: C15

DESCRIPTION: Perpendicular buttressed west tower with chamfered plinth. Rebuilt embattled parapet with crocketed finial; 3-light traceried bell-openings and some narrow slit openings. West front has restored 3-light window over doorway which has 4-centred arched head. Both have hood-moulds with animal and human figure stops. Stone bands have shields and figureheads. Figure gargoyles below parapet.

GRADE: II*

HERITAGE LIST NUMBER: 1115612

60 Barthomley
St Bertoline

LOCATION: Radway Green Road, Barthomley

POSTCODE: CW2 5PE

OS GRID REF: SJ767523

DATE: Late C15

DESCRIPTION: Square west tower with clasping buttresses, Gothic entrance with hood-mould and pair of 3 vertical panelled doors. The door opening is surmounted by a large 4-light, stained-glass Perpendicular window. There is a clock face facing north and 2-light Y-tracery windows to all faces at bell stage. The tower has a crenellated parapet with angle gargoyles and angle and intermediate crocketed pinnacles.

GRADE: I

HERITAGE LIST NUMBER: 1330063

61 Bidston
St Oswald

LOCATION: Bidston Village Road, Bidston

POSTCODE: CH43 7QT

OS GRID REF: SJ283903

DATE: Early C16

DESCRIPTION: 3-stage west tower with angle buttresses. Perpendicular stilted moulded arch to west door with drop-ended hood-mould over. Frieze of quatrefoil panels and coats of arms over doorway. 3-light window above with ogee hood-mould continuing as string-course. 3-light bell-chamber window in third stage. Embattled parapet with corbel-table.

GRADE: II

HERITAGE LIST NUMBER: 1201549

62 Bosley
St Mary the Virgin

LOCATION: Leek Road, Bosley

POSTCODE: SK11 0NX

OS GRID REF: SJ918655

DATE: c.1500

DESCRIPTION: Red sandstone tower of 3 stages has C19 boarded west door in medieval arch, diagonal buttresses with tooled quoins above, roughly squared rubble walls, west window of 2 trefoil lights, bell-openings converted to 3 trefoil lights 1878–79, hood-moulds on grotesque and head corbels. Top 2 metres and battlements are of 1878–79.

GRADE: II*

HERITAGE LIST NUMBER: 1138907

63 Brereton
St Oswald

LOCATION: Brereton Green, through lodge gates, 500 yards

POSTCODE: CW11 1RY

OS GRID REF: SJ781648

DATE: c.1500

DESCRIPTION: Embraced 3-stage west tower. Replaced door, 4-light panel-traceried west window, small rectangular opening to ringers' chamber, 2-light bell-opening in each face, crenellation and 4 pinnacles; square buttresses on west corners; one to north and south at east corners.

GRADE: II*

HERITAGE LIST NUMBER: 1229297

64 Bunbury
St Boniface

LOCATION: Wyche Road, Higher Bunbury

POSTCODE: CW6 9PN

OS GRID REF: SJ569581

DATE: C14

DESCRIPTION: Square west tower with C14 opening and curvilinear window above. There are lancets to 2 faces at second stage and, at bell stage, louvred Perpendicular windows. There are 3 times reducing buttresses to the quoins and a crenellated parapet with crocketed pinnacles.

GRADE: I

HERITAGE LIST NUMBER: 1138626

65 Cheadle St Mary

LOCATION: High Street, Cheadle

POSTCODE: SK8 1PX

OS GRID REF: SJ856886

DATE: 1520–40

DESCRIPTION: 3-stage tower with diagonal buttresses, moulded bands, clock in stage 2, 4-light belfry opening with hood-mould, castellated parapet and grotesque gargoyles. The west door and window are C19.

GRADE: I

HERITAGE LIST NUMBER: 1241643

66 Church Lowton All Saints

LOCATION: Church Lane, Liverpool Road West, Stoke-on-Trent

POSTCODE: ST7 3DE

OS GRID REF: SJ822557

DATE: Before 1555

DESCRIPTION: Perpendicular tower of before 1555. The western face has diagonal buttresses to the corners. Deeply chamfered doorway with pointed arch, hood-mould and label stops. 3-light window above probably C19, with 3 pointed lights. Staircase lancets to right of this. 2-light louvred belfry opening above with Y-tracery, hood-mould and figurehead label stops, that at right being a grotesque head. Battlemented parapet with piers to the corners. The north face of the tower has a C20 service lean-to attached to its lower stage and a string-course between the stages. Similar belfry to that on the western front. The southern face has the initials "I B" divided by a chalice in relief at right above the string-course and a circular clock face to the centre below the belfry opening.

GRADE: II*

HERITAGE LIST NUMBER: 1161743

67 Daresbury All Saints

LOCATION: Daresbury Lane, Daresbury

POSTCODE: WA4 4AE

OS GRID REF: SJ581828

DATE: C16

DESCRIPTION: Red sandstone tower has angle buttresses, a large west Perpendicular window, louvred windows to all faces at bell stage, gargoyles and crenellations.

GRADE: II*

HERITAGE LIST NUMBER: 1130450

68 Disley St Mary

LOCATION: Red Lane, Disley

POSTCODE: SK12 2NP

OS GRID REF: SJ974845

DATE: 1527–58

DESCRIPTION: West tower of 4 stages, with angle buttresses and castellated parapet showing the bases of 8 pinnacles. Surmounted by ornate wrought-iron weathervane moved from Stockport parish church. West window of 3 lights with straight mullions and 3-light louvred bell-opening with each light round-headed. Sundial on south face. Porch originally on south of nave, also castellated with crocketed pinnacles and a central cross. Entrance under slightly pitched plain lintel.

GRADE: II*

HERITAGE LIST NUMBER: 1231620

69 Frodsham St Laurence

LOCATION: Church Road, Frodsham

POSTCODE: WA6 6AD

OS GRID REF: SJ521772

DATE: C14

DESCRIPTION: 3-stage tower has low ramp roof to cellar vestry on west side; diagonal west and square east buttresses; 3-light panel-tracery west window; clock on north and south faces; 2-light bell-openings; crenellated parapet.

GRADE: I

HERITAGE LIST NUMBER: 1253193

70 Gawsworth
St James the Great

LOCATION: Church Lane, Gawsworth

POSTCODE: SK11 9RJ

OS GRID REF: SJ890697

DATE: c.1497–1536

DESCRIPTION: Red sandstone tower. West face: moulded plinth with diagonal buttresses. Central western doorway with 4-centred arch with quatrefoils and daggers to spandrels. Row of square panels above containing pointed quatrefoils. Cyma-moulded string-course above. 3-light window above with hood-mould and figurehead stops. Canopied niche above supported by angel bearing a shield. Crocket to canopy. At this same level are similar, smaller canopied niches to the outer faces of the diagonal buttresses, crockets to canopies. 3 shields in relief above central niche with circular clock face above them. String-course below belfry stage which has 2-light louvred openings with hood-moulds and figurehead label stops. Further row of square decorated panels below battlemented parapet which has gargoyles to centres and angles, above which rise panelled piers with crocketed pinnacles. South and north faces have shields and similar belfry stage and above.

GRADE: II

HERITAGE LIST NUMBER: 1139497

71 Grappenhall
St Wilfrid

LOCATION: Church Lane, Grappenhall

POSTCODE: WA4 3EP

OS GRID REF: SJ639863

DATE: 1525–39

DESCRIPTION: 3-stage tower has Tudor-arched west door, restored 4-light west window with panel tracery, diagonal west buttresses and square east buttresses, paired bell-openings with quatrefoil heads and crenellation.

GRADE: I

HERITAGE LIST NUMBER: 1139320

72 Great Budworth
St Mary & All Saints

LOCATION: Church Street, Great Budworth

POSTCODE: CW9 6HF

OS GRID REF: SJ664775

DATE: 1500–20

DESCRIPTION: 3-stage tower of 1500–20 has diagonal buttresses, octagonal south-west turret, replaced oak west door in ornamented archway surmounted by band with carved coats of arms, decayed carved panel to each side of door, Tudor-arched west window; band; small arched bell-ringer's window on north, west and south faces, eroded bas-relief panel on north and south faces and clock on west face; band; paired 2-light bell-openings with transoms and stone louvres; crenellation with 8 crocketed pinnacles.

GRADE: I

HERITAGE LIST NUMBER: 1139156

73 Handley
All Saints

LOCATION: Whitchurch Road, Handley

POSTCODE: CH3 9DT

OS GRID REF: SJ466579

DATE: 1512

DESCRIPTION: Tower has diagonal buttresses and 4-centred head to west door under label mould with carved head stops. 3-light west window breaks bond with carved heads and has a crocketed ogee hood. A tablet to either side reads: "Hoc ca' panile f'x'm' e' a' d'ni MCCCCXII t'p'e R'c'i raulinso, rectoris ac prooibus b'n'facto ribus ejusdem orate." Above are 2 empty niches. Similar 3-light louvred bell-openings and gargoyles support embattled parapet.

GRADE: II*

HERITAGE LIST NUMBER: 1230337

74 Holmes Chapel
St Luke

LOCATION: The Square, Holmes Chapel

POSTCODE: CW4 7NE

OS GRID REF: SJ762673

DATE: c.1430

DESCRIPTION: Square sandstone tower with buttresses which reduce 4 times in height. Gothic headed, ledged and boarded oak door, on large plain strap hinges, in coved and headed opening surmounted by a hood-mould. There is a 2-light Perpendicular window over the door and an empty niche with splayed reveals above this. There is a small 1-light window below the 2-light louvred window opening at bell stage. Diamond-shaped clock dials to north and south faces of the tower. Plain coved cornice with angle gargoyles and crenellated parapet.

GRADE: I

HERITAGE LIST NUMBER: 1231322

75 Ince St James

LOCATION: Pool Lane, Ince

POSTCODE: CH2 4NX

OS GRID REF: SJ450763

DATE: c.1485–93

DESCRIPTION: Tower built by Simon Ripley c.1485–93, raised 2 courses in 1854, with diagonal buttresses; 2-light traceried windows to upper storey, with blank niche in west wall below.

GRADE: II*

HERITAGE LIST NUMBER: 1138815

76 Little Budworth St Peter

LOCATION: Vicarage Lane, Little Budworth

POSTCODE: CW6 9BY

OS GRID REF: SJ598654

DATE: C15

DESCRIPTION: Perpendicular tower of 3 stages. Western front: projecting plinth with moulded top. Diagonal buttresses. Central doorway with moulded surround and hood-mould with worn figurehead label stops. C18 door with 8 raised and fielded panels. String-course above at level of sill of window to first stage which has a roll and casement-moulded surround and renewed glazing bars of c.1800 with 3 round-headed arched lights. String-course at level of springing of arch continuing around tower and across buttresses and set with a series of figure and animal heads. Canopied niche above this and string-course below belfry opening which is louvred and of 3 lights and showing fenestration of c.1800. String-course above with gargoyles to angles and centre. Battlemented parapet above with chamfered machicolations.

GRADE: II*

HERITAGE LIST NUMBER: 1139201

77 Lower Peover St Oswald

LOCATION: The Cobbles, Crown Lane, Lower Peover

POSTCODE: WA16 9PZ

OS GRID REF: SJ743741

DATE: Probably 1582

DESCRIPTION: 3-stage crenellated west tower probably 1582, but looks a little earlier. Diagonal buttresses, oak west door, Y-traceried 2-light west window, 3-light mullioned windows to bell-ringer's chamber, lozenge clock faces on north, west and south faces and Y-traceried 2-light bell-openings.

GRADE: I

HERITAGE LIST NUMBER: 1139104

78 Malpas St Oswald

LOCATION: Church Street, Malpas

POSTCODE: SY14 8PP

OS GRID REF: SJ486471

DATE: C14

DESCRIPTION: Massive 3-stage tower has diagonal west buttresses and angle east buttresses, south-east octagonal turret, bands above bell-chamber and at base of crenellated parapet with crocketed corner pinnacles, round-arched west door, C14 reticulated 5-light west window and simple 2-light reticulated bell-openings; an empty niche to each side of west window.

GRADE: I

HERITAGE LIST NUMBER: 1135959

79 Marbury St Michael

LOCATION: Church Lane, Marbury

POSTCODE: SY13 4LL

OS GRID REF: SJ560456

DATE: C16

DESCRIPTION: Perpendicular tower. Western end: diagonal buttresses with offsets. Projecting plinth with moulded top. Double doorway to ground level with double reveals and hood-mould with label stops. Above is a 3-light early Perpendicular window with trefoil-headed lights and chamfered reveal with hood-mould and label stops. String-course above carved with grotesque heads and above that a 3-light louvred belfry opening with cusped lights, hood-mould and label stops. Further string-course with grotesque heads below the level of the battlemented parapet which has gargoyles at the corners and to the centre. The parapet has panelled piers to the corners and centre with crocketed pinnacles. North face: blank below the level of the lower string-course similar to the western face above this. South face: similar save for a diamond-shaped clock board with circular clock face set below the belfry opening.

GRADE: II*

HERITAGE LIST NUMBER: 1225604

80 Mellor St Thomas

LOCATION: Church Road, Mellor

POSTCODE: SK6 5LX

OS GRID REF: SJ982889

DATE: C15

DESCRIPTION: 3-stage tower of C15 on C12 foundations. The tower has a C17 door surround within a C15 arch, diagonal weathered buttresses, quoins, a castellated parapet with enriched finials, C15 mullion and transom west window, 2-light belfry openings and clock faces.

GRADE: II

HERITAGE LIST NUMBER: 1241990

81 Middlewich St Michael & All Angels

LOCATION: Hightown, Middlewich

POSTCODE: CW10 9AN

OS GRID REF: SJ704662

DATE: c.1500

DESCRIPTION: Crenellated 3-stage tower has diagonal buttresses, west window to vestry with intersecting tracery and panel work, clock faces to west and east and 2-light bell-ringer's chamber window to north and south, paired 2-light bell-openings under single ogee hood-moulds, quatrefoil band below parapet.

GRADE: II*

HERITAGE LIST NUMBER: 1138795

82 Mobberley St Wilfrid

LOCATION: Church Lane, Mobberley

POSTCODE: WA16 7RA

OS GRID REF: SJ792802

DATE: 1533

DESCRIPTION: West tower of 1533 by Richard Platt, mason.

GRADE: I

HERITAGE LIST NUMBER: 1139554

83 Mottram in Longdendale St Michael

LOCATION: Warhill, Church Brow, Mottram in Longdendale

POSTCODE: SK14 6JL

OS GRID REF: SJ998953

DATE: c.1486

DESCRIPTION: 4-stage tower with angled weathered buttresses, stone bands, castellated parapet and crocketed corner finials. 3-light west window with Perpendicular tracery, clock face above and 2-light belfry openings. Stair-turret in one corner.

GRADE: II*

HERITAGE LIST NUMBER: 1356436

84 Neston
St Mary & St Helen

LOCATION: High Street, Town Centre, Neston

POSTCODE: CH64 6TZ

OS GRID REF: SJ291774

DATE: C14

DESCRIPTION: 4-stage C14 tower incorporating masonry from church of 1170, probably restored 1697. Upper stages heavily restored and substantially rebuilt in 1875. Tower has low weathered plinth and diagonal buttresses. Small pointed door to west with trefoil window above; first stage has sunken narrow windows with flat heads. The upper levels have louvred 2-light traceried openings. Date stone set in the south face gives 1697 restoration date. Crenellated parapet with 4 main face gargoyles.

GRADE: II*

HERITAGE LIST NUMBER: 1387671

85 Nether Alderley
St Mary

LOCATION: Church Lane, Nether Alderley

POSTCODE: SK10 4TW

OS GRID REF: SJ841761

DATE: Early C16

DESCRIPTION: 4-stage tower has diagonal buttresses and bands at each stage. Pointed-arch door-case with Victorian Decorated west window. Simple 2-light window in third stage and a pair of louvred 2-light Y-traceried bell-openings above. Gargoyles support an embattled parapet.

GRADE: I

HERITAGE LIST NUMBER: 1138849

86 Northwich (Witton) St Helen

LOCATION: Church Road, Northwich

POSTCODE: CW9 5PB

OS GRID REF: SJ664738

DATE: Early C16

DESCRIPTION: Crenellated 4-stage tower has diagonal buttresses, west door, basket-arched panel-traceried 4-light west window, 2-light bell-ringer's windows on north and south faces, empty niche on west face, clock with faces to all sides, paired panel-traceried 2-light bell-openings, band carved with foliage, faces and beasts at belfry floor and, simpler, above west window and below bell-ringer's windows; gargoyles.

GRADE: I

HERITAGE LIST NUMBER: 1329880

87 Pott Shrigley St Christopher

LOCATION: Shrigley Road, Pot Shrigley

POSTCODE: SK10 5RT

OS GRID REF: SJ944792

DATE: C15

DESCRIPTION: West tower of 4 stages with deeply set west doorway and 2-light, louvred bell-opening; castellated parapet and pinnacles.

GRADE: I

HERITAGE LIST NUMBER: 1277234

88 Prestbury St Peter

LOCATION: The Village, Prestbury

POSTCODE: SK10 4XY

OS GRID REF: SJ901769

DATE: c.1480

DESCRIPTION: 4-stage tower has clasping square buttresses. West door has label mould on carved heads. 3-light west window breaks a bond of interlacing. Clock in third stage and 2-light louvred bell-openings above. Another band of interlacing below embattled parapet with crocketed pinnacles.

GRADE: I

HERITAGE LIST NUMBER: 1221919

89 Shotwick St Michael

LOCATION: Shotwick Lane, Shotwick

POSTCODE: CH1 6HX

OS GRID REF: SJ336718

DATE: c.1500

DESCRIPTION: Tower has diagonal buttresses and bands at first and second floor with carved faces. 3-light west window and louvred bell-openings. Band with gargoyles carries embattled parapet.

GRADE: I

HERITAGE LIST NUMBER: 1145903

90 Tarvin
St Andrew

LOCATION: Church Street, Tarvin

POSTCODE: CH3 8EB

OS GRID REF: SJ491669

DATE: Late C15

DESCRIPTION: Tower has a moulded plinth, heavy clasping pilasters, moulded bands at each stage and a band of quatrefoils below the gargoyles and embattled parapet. 4-centred arched west door is recessed behind a moulded and rebated door-case crowned by a crocketed ogee head on angel corbels. Empty niches to either side with vaulted and crocketed heads. Restored 4-light west window, niche and corbels, Queen Victoria Jubilee clock, and pair of 2-light louvred bell-openings in successive stages.

GRADE: I

HERITAGE LIST NUMBER: 1330302

91 Tattenhall
St Alban

LOCATION: Church Bank, Tattenhall

POSTCODE: CH3 9QE

OS GRID REF: SJ486586

DATE: Early C16

DESCRIPTION: Tower has diagonal buttresses, cyma-moulded west door-case and a 3-light west window, breaking a band of carved heads, with a crocketed ogee label mould, flanked by pairs of initialled shields below, empty niches above. Similar 3-light louvred bell-openings and gargoyles support an embattled parapet.

GRADE: II*

HERITAGE LIST NUMBER: 1230254

92 Thornton-le-Moors St Mary

LOCATION: Thorton Green Lane, Thornton-le-Moors

POSTCODE: CH2 4JQ

OS GRID REF: SJ441745

DATE: C14

DESCRIPTION: C14 embattled tower, partly rebuilt, has stepped buttresses, 3 chamfered stone bands, 2-light bell-openings with 4-centred stopped hood-moulds. C14 traceried 2-light window on south wall.

GRADE: I

HERITAGE LIST NUMBER: 1330242

93 Tilston St Mary

LOCATION: Church Road, Tilston

POSTCODE: SY14 7HB

OS GRID REF: SJ457505

DATE: C15

DESCRIPTION: Embattled tower with stepped corner buttresses, low-pitched early Tudor west doorway with hood-mould. Blocked south doorway. Deeply recessed pointed window with reticulated tracery and bell-openings. Canopied niche over west doorway. Figure stops to hood-moulds, figure gargoyles.

GRADE: II*

HERITAGE LIST NUMBER: 1129940

94 Wallasey St Hilary

LOCATION: Claremont Road, Wallasey

POSTCODE: CH45 3NH

OS GRID REF: SJ297922

DATE: C13 with C16 upper stage

DESCRIPTION: 3-stage stone tower with diagonal buttresses. Lower part C13, upper part has date stone to south face: 1530. Blocked arches to north and east faces. Signs of lean-to roof to east face; north face has blocked 2-light window inserted in arch. West face has blocked window of 3 cusped lights. South face has tall blocked opening with inserted entrance with 4-centred head and date stone above. 3-light louvred bell-openings, drip-mould and embattled parapet with gargoyles. Arches formerly opened into church destroyed c.1850.

GRADE: II

HERITAGE LIST NUMBER: 1259875

95 Weaverham St Mary the Virgin

LOCATION: Church Street, Weaverham

POSTCODE: CW8 3NJ

OS GRID REF: SJ617742

DATE: Mid-C15

DESCRIPTION: The bold tower has diagonal buttresses, and bands at each floor. Band of cross motifs below parapet with gargoyles and an embattled parapet. Sharply pointed arch to the west door, restored 3-light west window and similar louvred bell-openings in the top stage.

GRADE: I

HERITAGE LIST NUMBER: 1287233

96 Wilmslow St Bartholomew

LOCATION: Church Street, Wilmslow

POSTCODE: SK9 1AW

OS GRID REF: SJ848814

DATE: Early C16

DESCRIPTION: 4-stage west tower has diagonal buttresses, moulded bands at first and third floor. Victorian west door and 3-light west window. Third stage has a quatrefoil in embrasure and above are 2-light louvred bell-openings. Embattled, pinnacled parapet is a Victorian copy.

GRADE: I

HERITAGE LIST NUMBER: 1222475

97 Winsford St Chad

LOCATION: Church Road, Winsford

POSTCODE: CW7 4DA

OS GRID REF: SJ650651

DATE: Early C16

DESCRIPTION: Tower of 3 storeys with diagonal buttresses and moulded plinth. Western front: arched doorway to ground floor with hood-mould and 3-light Perpendicular window above with hood-mould at level of string-course. Plain stage to middle with 3-light belfry openings to top with band of quatrefoils above and string-course above that with gargoyles above diagonal buttresses and at centre. Battlemented parapet: above. The north face has a 2-light C19 window to the middle stage with similar belfry stage as on west front.

GRADE: II*

HERITAGE LIST NUMBER: 1139180

98 Woodchurch Holy Cross

LOCATION: Church Lane, Woodchurch

POSTCODE: CH49 5LS

OS GRID REF: SJ276868

DATE: C14

DESCRIPTION: Tower refaced in later C17, but largely C14. 2 stages, the first stage very high. Massive stepped angle buttresses to south-west and north-west added in 1675 (dated). Square stair-turret to south-east. 3-light decorated window to bell-chamber with clock below. Embattled parapet with some head corbels below.

GRADE: II*

HERITAGE LIST NUMBER: 1217887

99 Wrenbury St Margaret

LOCATION: Nantwich Road, Wrenbury

POSTCODE: CW5 8HW

OS GRID REF: SJ594478

DATE: Early C16

DESCRIPTION: The western face has diagonal buttresses with offsets, doorway to the ground floor centre now blocked to its lower body and having a window with interlacing tracery above. Shields to the spandrels and a Tudor hood-mould. Above this is a pointed-arch window with deep chamfered surround which has C19 panel tracery inserted and a hood-mould with figurehead label stops. String-course above on which rests a central lancet with a belfry opening of 2 cusped lights with a quatrefoil above. Louvres to the openings and a hood-mould over. Gargoyle to left of centre and battlemented parapet above with crocketed pinnacles to the corners. The northern side is plain to the lower body with a similar arrangement of lancet and belfry openings above the string-course as appears on the western front. The southern face is essentially similar save that there is an octagonal staircase turret at right rising above the level of the tower parapet and finishing in an ogee lead cap with a weathercock. Above the string-course is a circular iron clock.

GRADE: II*

HERITAGE LIST NUMBER: 1357455

100 Wybunbury St Chad

LOCATION: Main Road, Wybunbury

POSTCODE: CW5 7LS

OS GRID REF: SJ700498

DATE: C15 or C16

DESCRIPTION: Square tower of 5 stages. Western front: moulded projecting plinth with chamfered top and string-courses between stages, all common to whole building. Clasping buttresses to lower body (probably early C19) dying to diagonal buttresses via broaches. Central doorway with richly moulded deep reveal (probably re-carved C19). Quatrefoils to spandrels and niches to either side with crocketed ogee heads each containing a sculpture of a bishop. Pair of decorated square panels above immediately below the brattished string-course. Central window above with 3-light interlacing tracery in casement-moulded surround of C19 date. Niches to either side with statues under projecting canopies with crocketed spirelets above. Brattished string-course above. Central niche to third stage similar to those at the second stage here holding a statue of a king with sword. The fourth stage has a central square wrought-iron clock face. The fifth stage has a pair of 2-light belfry openings under a common ogee hood-mould. Above is a row of decorated square panels. There are gargoyles to the angles and centre. The battlemented parapet has crocketed pinnacles to the corners. The north and south faces are similar save for the lack of niches and the door and that they each have a 2-light window to the third stage. The rear has a cambered archway originally giving access to the church, the ridge of which rose to the top of the second stage. The third stage has a lancet but no clock face. The tower lists to the north due to the sandy soil on which it is built. At the time of the resurvey (1985) it was in danger of collapse. In 1833 the earlier church was demolished and the tower straightened by James Trubshaw.

GRADE: II*

HERITAGE LIST NUMBER: 1279389

Diocese of
Durham

101 Auckland St Andrew

LOCATION: Crown Street, Bishop Auckland

POSTCODE: DL14 6RU

OS GRID REF: NZ217284

DATE: C13 with C15 top stage

DESCRIPTION: The 4-stage tower has shallow west buttress flanked by tall lancets; shouldered heads to first stage lights; third stage round heads to 2 shafted lights and plain spandrel recessed under round arch; corbel tale above. Tall pointed arches to 2-light belfry openings under battlemented parapet. South-west polygonal stair-turret with slit lights has stone coping set against belfry stage.

GRADE: I

HERITAGE LIST NUMBER: 1196458

102 Aycliffe St Andrew

LOCATION: Church Lane, Aycliffe village

POSTCODE: DL5 6JP

OS GRID REF: NZ283222

DATE: Tower of C10, top added late C14

DESCRIPTION: Buttressed, 4-stage tower with possibly Saxon sandstone masonry in lower 2 stages and limestone above (Saxon-type masonry continues to third stage within tower); blocked 2-light openings with shouldered heads to third stage; pointed openings with Y-tracery to belfry; embattled parapet. Upper stages of tower refaced c.1220.

GRADE: I

HERITAGE LIST NUMBER: 1322806

103 Billingham St Cuthbert

LOCATION: Church Lane, on Billingham Green

POSTCODE: TS23 1BW

OS GRID REF: NZ457223

DATE: C10

DESCRIPTION: Coursed rubble masonry battlemented tower of C10, top added late C14. Original openings in tower, 2-light with baluster shaft and pierced spandrel in bell stage.

GRADE: I

HERITAGE LIST NUMBER: 1139241

104 Brancepeth St Brandon

LOCATION: North Drive, adjacent to Brancepeth Castle

POSTCODE: DH7 8EA

OS GRID REF: NZ225377

DATE: Late C12

DESCRIPTION: Unbuttressed 4-stage tower with round-headed loops on second stage, paired lancets above and embattled parapet on corbel-table.

GRADE: I

HERITAGE LIST NUMBER: 1158956

105 Durham St Giles

LOCATION: Gilesgate, Durham

POSTCODE: DH1 1HS

OS GRID REF: NZ283427

DATE: Early C13 and early C15

DESCRIPTION: Square tower of 4 stages has clasping buttresses flanking restored 3-light west window with Perpendicular tracery; top stages have small square-headed window and 2-light belfry openings, traceried except on east side, below pinnacled battlements.

GRADE: I

HERITAGE LIST NUMBER: 1159991

106 Durham St Oswald

LOCATION: Church Street, Durham

POSTCODE: DH1 3DQ

OS GRID REF: NZ276419

DATE: C15

DESCRIPTION: The 4-stage tower has renewed 4-light west window and blank north and south first stage; small 2-light windows above with cusped ogee heads; very small square windows in third stage; tall 2-light belfry openings with 2-centred heads. Diagonal buttresses with offsets; battlemented parapet with corner gargoyles and pinnacles.

GRADE: II*

HERITAGE LIST NUMBER: 1120678

107 Easington
St Mary

LOCATION: Hall Walks, Easington village

POSTCODE: SR8 3GA

OS GRID REF: NZ414435

DATE: C11

DESCRIPTION: Romanesque 4-stage west tower defined by bands. Fragment of C10–11 grave slab in base of west wall. Romanesque masonry up to and including corbel-table. 2 round-arched openings and later battlemented parapet. 2 massive C14 diagonal buttresses.

GRADE: I

HERITAGE LIST NUMBER: 1231813

108 Egglescliffe
St John the Baptist

LOCATION: Church Road, Egglescliffe

POSTCODE: TS16 9DT

OS GRID REF: NZ421133

DATE: C14

DESCRIPTION: Mainly Perpendicular church on site of earlier Saxon church. Ashlar with battlemented west tower.

GRADE: I

HERITAGE LIST NUMBER: 1104901

109 Gainford St Mary

LOCATION: Low Green, Gainford

POSTCODE: DL2 3DW

OS GRID REF: NZ169167

DATE: C13, extensively rebuilt in C19

DESCRIPTION: 3-stage west tower with C18 stair-turret. Belfry has corbelled parapet with 4 corner pinnacles of 1786.

GRADE: I

HERITAGE LIST NUMBER: 1121114

110 Hart St Mary Magdalene

LOCATION: Magdalene Drive, Hart village

POSTCODE: TS27 3BS

OS GRID REF: NZ471351

DATE: C12

DESCRIPTION: Tower of 3 stages with chamfered step-backs has c.1800 straight parapet stepped out at eaves and ogee angle finials. Louvred rectangular belfry openings to west and south faces; other tower openings are lancets.

GRADE: I

HERITAGE LIST NUMBER: 1249898

111 Hartlepool All Saints

LOCATION: Stranton Green, Stranton

POSTCODE: TS24 7LD

OS GRID REF: NZ509320

DATE: Late C13

DESCRIPTION: West tower of 3 stages defined by chamfered step-backs; diagonal buttresses with offsets; embattled parapet and angle and intermediate pinnacles. 3 grouped lancets to west face of lower stage; slit lights to south and west faces of middle stage; upper stage has clock faces below round-headed openings holding paired trefoil-headed lights.

GRADE: II*

HERITAGE LIST NUMBER: 1250534

112 Haughton-le-Skerne St Andrew

LOCATION: Haughton Road, Haughton Green

POSTCODE: DL1 2TZ

OS GRID REF: NZ308159

DATE: C12

DESCRIPTION: Norman west tower of 2 stages built on site of earlier Saxon church.

GRADE: I

HERITAGE LIST NUMBER: 1160229

113 Heighington St Michael

LOCATION: East Green, Church View, Heighington

POSTCODE: DL5 6PN

OS GRID REF: NZ249223

DATE: Saxon

DESCRIPTION: Saxon 3-stage tower: probably C19 south doorway; small round-headed windows on south and west; twin bell-openings on third stage; embattled parapet with gargoyles.

GRADE: I

HERITAGE LIST NUMBER: 1322953

114 Kelloe St Helen

LOCATION: Sharon Avenue, Kelloe

POSTCODE: DH6 4PT

OS GRID REF: NZ345365

DATE: C11

DESCRIPTION: 3-stage tower with massive stepped buttress and C18 diamond sundial on south and small round-headed belfry windows on each face.

GRADE: I

HERITAGE LIST NUMBER: 1120748

115 Lanchester All Saints

LOCATION: Durham Road, Lanchester

POSTCODE: DH7 0NG

OS GRID REF: NZ168474

DATE: C14

DESCRIPTION: Tower has short diagonal buttresses flanking a chamfered window with cusped Y-tracery; clock in moulded surround at top of high first stage; belfry has C19 openings in similar style to west window, with polygonal corner pilasters, under parapet string with stone drains below battlements with corner pinnacles. Small rectangular lights in north and south elevations and in stair-turret on south-east.

GRADE: I

HERITAGE LIST NUMBER: 1185983

116 Monkwearmouth St Peter with St Cuthbert

LOCATION: St Peter's Way, Monkwearmouth

POSTCODE: SR6 0DY

OS GRID REF: NZ401577

DATE: Late C7–late C10

DESCRIPTION: The 5-stage tower has open-arched porch with 3 round-headed arches on impost blocks, the west arch with balusters supporting eroded low-relief carving of interlace-type animals. Next stage has single round-headed opening; above this, in former gable, an eroded figure; fourth stage a single plain arched light; belfry has 2 arched lights with baluster mullion; small blocked roundel above. The lower stages of the tower date from 674 to c.700, the upper stages of the tower from before 1000.

GRADE: I

HERITAGE LIST NUMBER: 1217958

117 Pittington St Laurence

LOCATION: Hallgarth, High Pittington

POSTCODE: DH6 1AB

OS GRID REF: NZ328436

DATE: Late C12/early C13, with C15 belfry stage

DESCRIPTION: The tower has a long west lancet under drip-mould; small lancet in set-back second stage; clock above; paired cusped lights in set-back belfry are deeply recessed in round-headed openings, under embattled parapet with corner pinnacles.

GRADE: I

HERITAGE LIST NUMBER: 1310892

118 Redmarshall St Cuthbert

LOCATION: Church Lane, west end of village

POSTCODE: TS21 1EU

OS GRID REF: NZ386212

DATE: Norman

DESCRIPTION: Unbuttressed west tower of roughly coursed rubble. Mainly Norman with battlements of C15.

GRADE: I

HERITAGE LIST NUMBER: 1140001

119 Sedgefield St Edmund

LOCATION: Cross Hill, Sedgefield

POSTCODE: TS21 3AT

OS GRID REF: NZ356288

DATE: c.1490

DESCRIPTION: Diagonally buttressed, 3-stage tower has stair-turret on south wall; square lights to second stage; louvred, 3-centred belfry openings; embattled parapet with octagonal turrets and spirelets.

GRADE: I

HERITAGE LIST NUMBER: 1121482

120 Staindrop St Mary

LOCATION: Front Street, Staindrop

POSTCODE: DL2 3NL

OS GRID REF: NZ131206

DATE: Early C12

DESCRIPTION: 3-stage tower with 2-centred arched door in west stair-turret; lancet in high first stage; paired lancets in short second stage; renewed Y-tracery in corbelled-out belfry stage. Buttresses, clasping at corners; parapets with chamfered coping except for roll-moulded south aisle and embattled tower.

GRADE: I

HERITAGE LIST NUMBER: 1338594

121 Stanhope St Thomas

LOCATION: Front Street, north of Market Square

POSTCODE: DL13 2UW

OS GRID REF: NZ997392

DATE: C12–C13

DESCRIPTION: Wide 3-stage tower has small round-headed window in first stage, painted wooden clock above, and paired chamfered belfry lancets with central shaft. Corbel-table supports battlements of grey stone, perhaps C17, of millstone grit, with prominent stone drains, sloped coping and small eroded corner pinnacles.

GRADE: II*

HERITAGE LIST NUMBER: 1231701

122 Whickham St Mary

LOCATION: Church Chare, village centre, Whickham

POSTCODE: NE16 4SL

OS GRID REF: NZ210613

DATE: C14

DESCRIPTION: West tower with north stair-turret; 3-light west window; upper stages have lancet windows beneath paired shouldered bell-openings and battlemented parapet; clock dial on south.

GRADE: I

HERITAGE LIST NUMBER: 1185359

123 Wolsingham St Mary & St Stephen

LOCATION: Church Lane, Wolsingham

POSTCODE: DL13 3AW

OS GRID REF: NZ074374

DATE: Norman

DESCRIPTION: Tower, Norman to nave roof level, of 3 stages with set-backs; small round-headed stair-turret lights in west front; battlemented top stage has paired belfry openings under 2-centred arched paired drip-moulds.

GRADE: II

HERITAGE LIST NUMBER: 1232841

Diocese of

Leeds

124 Ackworth St Cuthbert

LOCATION: Purston Lane, High Ackworth

POSTCODE: WF7 7ED

OS GRID REF: SE440180

DATE: C15

DESCRIPTION: 4-stage west tower with diagonal buttresses and weathered plinth has a wide 2-centred arched west doorway with 2 orders of cavetto moulding and C19 hood-mould, above this a restored 4-centred arched 3-light window with matching surround, Perpendicular tracery, and hood-mould, a recessed lancet to the third stage, a drip-band carried round; in each side of the top stage, a recessed belfry window of 2 cusped lights with a quatrefoil in the head and a hood-mould, a drip-band with 3 gargoyles, a restored embattled parapet with crocketed corner pinnacles; and a flag mast rising from the centre of the roof; the south side has a niche to the third stage, and a diamond-shaped clock face above this (the band stepped over it).

GRADE: II*

HERITAGE LIST NUMBER: 1313257

125 Ainderby Steeple St Helen

LOCATION: Northallerton Road, Ainderby Steeple

POSTCODE: DL7 9PZ

OS GRID REF: SE334921

DATE: C15

DESCRIPTION: 3-stage tower clasped by aisles. Diagonal offset buttresses to west end and 3-light cusped arched west window with hood-mould. Small cusped 1-light opening above each aisle. Clock faces to west and south. A 2-light arched cusped belfry window with hood-mould to each face. Band. Embattled parapet.

GRADE: I

HERITAGE LIST NUMBER: 1315088

126 Aldborough St Andrew

LOCATION: Aldborough, Boroughbridge

POSTCODE: YO51 9ES

OS GRID REF: SE406664

DATE: C15

DESCRIPTION: Low battlemented west tower.

GRADE: I

HERITAGE LIST NUMBER: 1173835

127 Almondbury All Hallows

LOCATION: Westgate, Almondbury

POSTCODE: HD5 8XF

OS GRID REF: SE168151

DATE: C15

DESCRIPTION: West tower apparently C15, possibly the result of an indulgence of 1485 for the repair of the church. Tower has diagonal buttresses with many set-offs and gargoyles at top. Crenellated parapet. Crocketed pinnacles. West door in deeply moulded 2-centred arch. 3-light west window with Perpendicular tracery, in 2-centred arch with hood-mould. Large 3-light bell-openings in 2-centred openings with hood-mould.

GRADE: I

HERITAGE LIST NUMBER: 1225096

128 Arncliffe St Oswald

LOCATION: Arncliffe

POSTCODE: BD23 5QB

OS GRID REF: SD933720

DATE: C15

DESCRIPTION: West tower of 2 stages with offset diagonal buttresses; a 3-light restored Perpendicular west window, clock to south side; belfry windows of 2 trefoil-headed lights under a continuous hood-mould; restored embattled parapet with crocketed corner finials.

GRADE: II

HERITAGE LIST NUMBER: 1132236

129 Askrigg St Oswald

LOCATION: On Market Place, Askrigg

POSTCODE: DL8 3HT

OS GRID REF: SD948910

DATE: C15

DESCRIPTION: Tower of 3 storeys. West face: ground-floor 2-light cinquefoiled window with Perpendicular tracery in straight-sided pointed arch with hood-mould; first-floor, 1-light trefoiled window. 2-light belfry openings with trefoil heads and hood-moulds; crenellated parapet and corner pinnacles.

GRADE: I

HERITAGE LIST NUMBER: 1301613

130 Badsworth St Mary

LOCATION: New Road, Badsworth

POSTCODE: WF9 1AJ

OS GRID REF: SE463150

DATE: C15

DESCRIPTION: West tower, nave with north and south aisles, chancel with added north vestry. 4-stage tower has diagonal buttresses, weathered plinth, weathered bands on 3 levels; 2-centred arched west doorway with 3 orders of moulding, and hood-mould; large 2-centred arched west window of 3 lights with Perpendicular tracery, hollow-moulded reveal, and hood-mould; above this, a small crocketed niche; 2-light belfry windows with transoms and cinquefoil lights with wooden louvres, hollow-moulded surrounds and hood-moulds with figured stops; restored embattled parapet with crocketed corner pinnacles; clock face on south side of third stage.

GRADE: I

HERITAGE LIST NUMBER: 1052219

131 Bardsey
All Hallows

LOCATION: Church Lane, Bardsey

POSTCODE: LS17 9DN

OS GRID REF: SE365431

DATE: C9–C10

DESCRIPTION: Anglo-Saxon tower 850–950 raised later C10, parapet early C20. Slender tower embraced by aisles gives the west front an unusual appearance as if the tower were rising out of a gable. Long and short work on this face: 8 courses of large dressed quoins; masonry disturbed from insertion of 2-light Perpendicular window matching flanking windows set in west end of aisles with C12 lancets set between tower; earlier roofline of aisles clearly discernible as is steeply pitched gable on tower, thought to indicate the presence of an original Anglo-Saxon porch. West and north faces of tower have a small light under lancet and inserted 2-light Perpendicular belfry windows with clock face to north; south face has pair of C12 2-light baluster belfry windows; east face has traces of herringbone masonry; embattled parapet with corbel-table (restored).

GRADE: I

HERITAGE LIST NUMBER: 1135652

132 Barnoldswick St Mary-le-Ghyll

LOCATION: Ghyll Lane, Barnoldswick

POSTCODE: BB18 6JG

OS GRID REF: SD893480

DATE: Early C16 (tower said to be dated 1524)

DESCRIPTION: Tower is rectangular and simply treated with gargoyle string-course and embattled parapet. Diagonal buttresses and square stair-turret at south-east corner. West window with arched head of 3 lights, and belfry openings of 2 lights with louvres, also arched, all with hood-moulds.

GRADE: I

HERITAGE LIST NUMBER: 1073421

133 Barwick in Elmet All Saints

LOCATION: Aberford Road, Barwick in Elmet

POSTCODE: LS15 4EA

OS GRID REF: SE401374

DATE: C15

DESCRIPTION: 2-stage embattled tower. West face has pointed-arch doorway of 2 chamfered orders under hood-mould; 3-light window above has panel tracery, hood-mould with carved-face stops; above left, niche with sculpture of the benefactor Sir Henry Vavasour holding in his hands a block of stone, inscription under has date 1455; another niche above to right (vacant) has inscription of Richard Burnham; large clock face below second-stage band; 2-light belfry window, parapet carried on corbel-table. South face: 5 small arrow slits one above the other to light stair.

GRADE: II*

HERITAGE LIST NUMBER: 1357161

134 Batley All Saints

LOCATION: Stocks Lane, Batley

POSTCODE: WF17 8PA

OS GRID REF: SE242245

DATE: c.1485

DESCRIPTION: Low Perpendicular west tower. Corbelled-out, machicolated and castellated parapet with tall square pinnacles. 2-light louvred and traceried openings to bell-chamber.

GRADE: I

HERITAGE LIST NUMBER: 1134620

135 Bedale St Gregory

LOCATION: North End, Bedale

POSTCODE: DL8 1AF

OS GRID REF: SE265884

DATE: C13–C15

DESCRIPTION: 5-stage tower. Decorated style with Perpendicular-style bell stage. Offset angle buttresses, stair-turret with 5 chamfered slit openings. To left, three 2-light cusped, arched chamfered windows with hood-moulds. Stone band and clock face between second and third windows. Above, a pair of 2-light flat-headed belfry windows with cusped lights and hood-moulds. Small pilaster buttress between and either side. Band. Diagonally set gargoyles at corners. Embattled parapet with pinnacles at each corner and centre of each side. Panelled west door in doorway of 2 chamfered orders under hood-mould with head stops. Above, a 3-light window with Decorated tracery and hood-moulds with head stops.

GRADE: I

HERITAGE LIST NUMBER: 1151205

136 Bentham
St John the Baptist

LOCATION: Wennington Road, Bentham

POSTCODE: LA2 7DB

OS GRID REF: SD648689

DATE: C15

DESCRIPTION: 3-stage Perpendicular west tower. Restored 2-centred head to west door, dripstone; 1870s 2-leaf plank doors with decorative wrought-iron strap hinges. Second stage: 1870s 3-light Perpendicular window with 2-centred heads and curved dripstone. Bell stage: 2-light trefoil-headed window with mouchettes in spandrels; drip-mould. Diagonal buttresses; partly embattled parapet with 4 crocketed finials all c.1980 (2 C15 finials in churchyard, west of tower).

GRADE: I

HERITAGE LIST NUMBER: 1157613

137 Bingley
All Saints

LOCATION: Old Main Street, Bingley

POSTCODE: BD16 2RH

OS GRID REF: SE105394

DATE: Late C15

DESCRIPTION: Embattled tower with offset diagonal buttresses, clock and sundial. Pointed-arch west door set in base of tower with 3-light traceried window above and 2-light window below. Top stage, C19 with 2-light belfry openings. Buttresses rise into crockets.

GRADE: II*

HERITAGE LIST NUMBER: 1200095

138 Birstall St Peter

LOCATION: Kirkgate, Birstall

POSTCODE: WF17 9PB

OS GRID REF: SE218262

DATE: C11 and C15

DESCRIPTION: The tower is of 3 stages, Norman in its lower 2 stages, to which diagonal buttresses at the bell stage, and an embattled parapet on a corbel-table, with stout corner pinnacles, were added C15. The lower stage has a narrow west door, of C19 detail but earlier proportions, which has relief foliage in the spandrels, below a 2-light Perpendicular west window. The second stage has narrow round-headed north and south windows, above which is a round clock on the west face. The bell stage has 2-light transomed Perpendicular openings, below which is a 1660 sundial on the south face.

GRADE: II*

HERITAGE LIST NUMBER: 1134648

139 Bolton St Oswald

LOCATION: West end of Castle Bolton

POSTCODE: DL8 4ET

OS GRID REF: SE033919

DATE: Late C14

DESCRIPTION: 2-storey west tower, quoined, with 1-light, pointed-arch west window under label and, above, a 2-light belfry opening without mullion, with trefoiled heads to monolithic lintel; single lancet to south.

GRADE: II*

HERITAGE LIST NUMBER: 1130886

140 Bolton on Swale St Mary

LOCATION: Visible from B6271

POSTCODE: DL10 6AQ

OS GRID REF: SE252991

DATE: Late C16

DESCRIPTION: Tower of 3 stages. Diagonal buttresses to west with carving to offsets. Stair-turret in south-east corner with light vents. Ground-floor west window: 3 trefoil-headed lights below semicircular head with tracery in spandrels; label with head stops; double-chamfered jambs. Light vents to north, west and south of ringing chamber, clock to west. Belfry openings: crossed cinquefoil-headed lights with continuous decorated lintels, with hood-moulds. Ashlar battlements with coats of arms.

GRADE: II*

HERITAGE LIST NUMBER: 1131521

141 Bracewell St Michael

LOCATION: Gisburn Road, Bracewell

POSTCODE: BD23 3JU

OS GRID REF: SD863485

DATE: C15 or early C16

DESCRIPTION: West tower very simple, in 2 stages marked only by a pronounced set-back, probably indicating the early date of the lower stage which, however, has only flat-headed openings, the west window being double-chamfered of 2 lights. The slimmer upper stage, probably C15, has rough 2-light louvred belfry windows, corbelled embattled parapet and waterspouts. On its eastern face is the scar of a steep roof.

GRADE: I

HERITAGE LIST NUMBER: 1259162

142 Bradford
Cathedral Church of St Peter

LOCATION: Stott Hill, east of city centre, Bradford

POSTCODE: BD1 4EH

OS GRID REF: SD167333

DATE: Late C15

DESCRIPTION: West tower with heavy Perpendicular detailing of 1493–1508. 6-light window, twin 2-light bell-openings, battlements, 8 pinnacles.

GRADE: I

HERITAGE LIST NUMBER: 1133250

143 Broughton
All Saints

LOCATION: Church Lane, Broughton

POSTCODE: BD23 3AN

OS GRID REF: SD934504

DATE: Early C16

DESCRIPTION: Tower has diagonal buttresses, gargoyles, battlements and pinnacles. Arched belfry windows and 2-light square-headed window to first stage.

GRADE: I

HERITAGE LIST NUMBER: 1132291

144 Burnsall
St Wilfrid

LOCATION: B6160, Burnsall

POSTCODE: BD23 6BN

OS GRID REF: SE032615

DATE: C16

DESCRIPTION: Perpendicular 3-stage tower: offset diagonal buttresses. Large arched Perpendicular west window, cusped of 3 lights with hood-mould, to right a board door in 4-centred arch surround forming a buttressed porch with small 1-light window above. Second stage has an octagonal clock face to west, to south a Perpendicular segmental-arched 2-light window. Chamfered band. Arched belfry windows, cusped of 3 lights, hood-mould. Moulded band. Embattled parapet with 4 small pinnacles to each corner.

GRADE: I

HERITAGE LIST NUMBER: 1131740

145 Calverley
St Wilfrid

LOCATION: Town Gate, Calverley

POSTCODE: LS28 5NF

OS GRID REF: SE208372

DATE: C14 and C15

DESCRIPTION: C14 tower with C15 upper stage; plinth, angle buttresses and south-east corner stair-tower with 3 lancets, set in base is doorway with ogee lintel. West window of 3 cusped lights surmounted by quatrefoil. Set above in all faces is a 2-light arched window with cusped lights, that to south blocked by octagonal clock face. Cyma-moulded band separates stages; top stage has 3-light belfry openings with panel tracery; Lombard frieze to embattled parapet surmounted by crocketed pinnacles.

GRADE: II*

HERITAGE LIST NUMBER: 1226034

146 Catterick St Anne

LOCATION: High Green, Catterick

POSTCODE: DL10 7LL

OS GRID REF: SE240980

DATE: C15

DESCRIPTION: Perpendicular-style west tower of 3 stages. Stair-turret with light vents in south-east corner. 3-light west window. Light vents to west and south sides of ringing chamber with clock. 2-light belfry openings. Crenellated parapet with corner finials.

GRADE: I

HERITAGE LIST NUMBER: 1131488

147 Cawthorne All Saints

LOCATION: Church Lane, Cawthorne

POSTCODE: S75 4HL

OS GRID REF: SE285080

DATE: C15

DESCRIPTION: The Perpendicular tower of 2 tall stages was diagonal buttresses, a west door, restored Perpendicular west window of 3 lights and 2-over-2-light bell-chamber openings. Restored crenellated parapet with gargoyles and tall corner pinnacles.

GRADE: II*

HERITAGE LIST NUMBER: 1191523

148 Clapham St James

LOCATION: Church Avenue, Clapham

POSTCODE: LA2 8DU

OS GRID REF: SD746694

DATE: C15

DESCRIPTION: West tower of 3 stages; west entrance has Tudor head, moulded surround, c.1900 2-leaf plank doors, hood-moulds. 3-light window immediately above has pointed arch, cavetto surround and Perpendicular tracery. Small trefoil-headed window in second stage with cavetto surround. In bell stage is deeply chamfered belfry opening with tracery now gone on west side, but 2-light trefoil-headed on north and south. Embattled parapet and central gargoyle on each face. Diagonal buttresses. Remains of former nave wall to north and south, that on south having chamfered lancet window.

GRADE: II

HERITAGE LIST NUMBER: 1132398

149 Collingham St Oswald

LOCATION: Church Lane, Collingham

POSTCODE: LS22 5BD

OS GRID REF: SE390461

DATE: C15

DESCRIPTION: Tower of 2 stages, chamfered plinth, moulded band. 3-light west window with arched lights and panel tracery, above is limestone block with shield in relief; moulded band; 2-light belfry windows set in square double-chamfered surrounds have arched lights with sunken spandrels; south face has clock face above small chamfered light. Embattled parapet with projecting rainwater chutes with gargoyles; crocketed corner pinnacles.

GRADE: II*

HERITAGE LIST NUMBER: 1225846

150 Cowthorpe St Michael

LOCATION: Oak Road, Cowthorpe

POSTCODE: LS22 5EZ

OS GRID REF: SE427528

DATE: C15

DESCRIPTION: Tower built partly in front of the west wall of the nave on 2 external buttresses with large central deep arch on 2 transverse ribs forming a recess in which there is a large pointed 3-light mullion and transom window with cusped lights, to nave wall. Second stage, south side: a circular cusped window, a string-course above. The bell stage openings have 4-centred arch lights in chamfered flat-headed surrounds, of 2 lights on each side except to east which is of 3 lights with shield stops to the hood-moulds. Oversailing battlemented parapet.

GRADE: I

HERITAGE LIST NUMBER: 1315417

151 Croft-on-Tees St Peter

LOCATION: The Mill Race, Croft-on-Tees

POSTCODE: DL2 2SG

OS GRID REF: NZ289098

DATE: C12–C13 and C19

DESCRIPTION: Tower of 2 stages, the lower C12–C13 and of red sandstone, the upper C19, of brown sandstone; stepped diagonal buttress; ground-floor pointed-arch window of two 5-cusped lights; set into the later stonework below the string, carved heads and a stone carved with "IHS", also a sundial; above the string a clock, and at each corner a coat of arms formerly recorded as being on the porch, that to left of Roland Place, that to right of Richard Clervaux (2 rival local landowners whose quarrels were settled by arbitration of Richard, Duke of Gloucester in 1478); belfry opening of 2 triple-cusped lights with pierced tracery; C19 battlemented parapet; matching belfry openings to east, north and west; also to west on lower stage trefoiled lancet window.

GRADE: I

HERITAGE LIST NUMBER: 1301945

152 Danby Wiske No dedication

LOCATION: Mounstrall Lane, Danby Wiske

POSTCODE: DL7 0LY

OS GRID REF: SE338983

DATE: C15

DESCRIPTION: Tower of ashlar with offset diagonal buttresses, stair-tower with 2 slit windows, 2-light C15 belfry window and embattled parapet.

GRADE: I

HERITAGE LIST NUMBER: 1150204

153 Darrington St Luke & All Saints

LOCATION: Church Lane, Darrington

POSTCODE: WF8 3AU

OS GRID REF: SE485202

DATE: C11 with C15 additions

DESCRIPTION: Norman tower. The stout, square, 3-stage west tower, embraced by the aisles up to the second stage, has a moulded 4-centred arched west doorway and a Perpendicular 3-light window above; at the second stage, the south side has a clock face and the north side has a small Norman window of 2 round-headed lights (now glazed) with a colonnette; the third stage, which is set back, has Norman belfry windows in the south and east sides, each of 2 round-headed louvred lights with a colonnette, an altered or inserted belfry window of 2 pointed lights in the west side; and an embattled parapet with restored corner pinnacles.

GRADE: I

HERITAGE LIST NUMBER: 1313210

154 Darton All Saints

LOCATION: Church Lane, Darton

POSTCODE: S75 5NQ

OS GRID REF: SE311099

DATE: C16

DESCRIPTION: Tall, Perpendicular-style, 2-stage tower of smoother ashlar with diagonal buttresses. Moulded west doorway, 3-light west window. Tall, transomed, 2-light bell-chamber openings. Crenellated parapet.

GRADE: I

HERITAGE LIST NUMBER: 1286515

155 Elland
St Mary the Virgin

LOCATION: The Cross, Church Street, Elland

POSTCODE: HX5 0RU

OS GRID REF: SE108522

DATE: c.1490

DESCRIPTION: Tower of 3 stages with diagonal buttresses and embattled parapet. West window of 3 lights with simple tracery. The belfry windows are of 2 lights.

GRADE: I

HERITAGE LIST NUMBER: 1184393

156 Emley
St Michael the Archangel

LOCATION: Church Street, Emley

POSTCODE: HD8 9TH

OS GRID REF: SE245133

DATE: C15

DESCRIPTION: 2-stage square west tower. Diagonal buttresses to tower, reducing with offsets. West door with moulded surround and hood-mould. Large 3-light traceried west window (restored). 4-light mullioned-and-transomed bell-chamber openings with traceried heads. Gargoyles, crenellated parapet and 8 crocketed pinnacles.

GRADE: I

HERITAGE LIST NUMBER: 1135290

157 Eryholme St Mary the Virgin

LOCATION: Eryholme, Darlington

POSTCODE: DL2 2PE

OS GRID REF: NZ320090

DATE: C13 with C16 additions

DESCRIPTION: Very low tower of 2 stages, and very small in plan, without plinth or buttresses, probably mainly C13; quoins; on south, west and north sides a C16 round-arched chamfered belfry opening with stone slate louvres; on west side a small chamfered ground-floor lancet and waterspout above the belfry opening; C16 battlemented parapet.

GRADE: II*

HERITAGE LIST NUMBER: 1131336

158 Farnham St Oswald

LOCATION: Stang Lane, Farnham

POSTCODE: HG5 9JD

OS GRID REF: SE348605

DATE: c.1500

DESCRIPTION: The tower has a belfry stage above the nave roofline with 2-light windows recessed in square chamfered surround. Plain parapet with stone spouts and corner pinnacles.

GRADE: I

HERITAGE LIST NUMBER: 1150301

159 Felkirk St Peter

LOCATION: Slack Lane, Felkirk

POSTCODE: S72 9DH

OS GRID REF: SE387126

DATE: C15

DESCRIPTION: Perpendicular tower of rectangular plan and 3 tall stages, with diagonal buttresses to half-height chamfered plinth and drip-bands on 2 levels, has 2-centred arched west doorway with studded plank door, 2 orders of hollow moulding, hood-mould with figured stops and apex shield (blank); large 2-centred arched west window (restored) of 4 cusped lights with Perpendicular tracery in the head; small inserted window to second stage; 2-centred arched belfry windows with 2 stages of cusped lights, traceried heads, and hood-moulds; large carved gargoyles at the corners and small ones in the centre; and restored embattled parapet with crocketed pinnacles at the corners and in the centre of each side.

GRADE: I

HERITAGE LIST NUMBER: 1265442

160 Forcett
St Cuthbert

LOCATION: B6274, Forcett

POSTCODE: DL11 7SD

OS GRID REF: NZ175123

DATE: C13

DESCRIPTION: Tower of 3½ stages, the lowest being relatively near the ground, forming a kind of basement; plinth; quoins; round-arched vents on ground and first stages; belfry opening of 2 segmental pointed-arch lights under monolithic lintel, divided by shaft with small volutes on capital; battlemented parapets; lancet window with hood-mould on ground floor of west side and matching belfry opening; square-plan stair-turret in north-east corner and matching belfry opening on north side; octagonal clock in moulded frame on east side.

GRADE: II

HERITAGE LIST NUMBER: 1131962

161 Gargrave
St Andrew

LOCATION: Church Street, Gargrave

POSTCODE: BD23 3NE

OS GRID REF: SD932539

DATE: C16

DESCRIPTION: West tower has diagonal buttresses, battlements and 8 pinnacles. 3-light windows including that to west, which is surmounted by an empty niche with cresting, below which is an inscription.

GRADE: II

HERITAGE LIST NUMBER: 1132086

162 Giggleswick St Alkelda

LOCATION: Church Street, Giggleswick

POSTCODE: BD24 0BE

OS GRID REF: SD812641

DATE: Mid-C15

DESCRIPTION: West entrance has basket arch and hood-mould; trefoil heads to 3-light window above, rectilinear tracery. Second stage string-course and 1-light chamfered window with trefoil head. Deeply chamfered trefoil-headed louvred window to bell stage. Gargoyle; embattled parapet; crocketed finials. Diagonal buttresses and projecting stair-turret on south side; clock face on east side.

GRADE: I

HERITAGE LIST NUMBER: 1157303

163 Gilling St Agatha

LOCATION: Gilling West, Richmond

POSTCODE: DL10 5JN

OS GRID REF: NZ182051

DATE: Late C11 and C15

DESCRIPTION: Late C11 tower raised in late C15, with stair-turret added in 1845; 3 stages with chamfered bands in between; quoins; 5-sided stair-turret giving access to ringing chamber, and with pyramidal stone roof; above, chamfered vesica window; on third stage, blocked round-arched window, and above that straight-headed double-chamfered belfry opening, formerly with 3 closely spaced mullions; crenellated parapet; central weathercock. On west side of tower, ground-floor segmental-arched hollow-moulded window with 3 pointed lights and label; second stage blank; third stage as south. North side of tower is blank except for third stage as south. East side of tower: octagonal wooden clock face on second stage; third stage as south.

GRADE: I

HERITAGE LIST NUMBER: 1316927

164 Goldsborough St Mary

LOCATION: Church Street, Goldsborough, Knaresborough

POSTCODE: HG5 8NR

OS GRID REF: SE384561

DATE: C14–C15

DESCRIPTION: 3-stage tower with moulded plinth, narrow window and shields to second stage, and Perpendicular belfry openings. Buttresses set back to lowest stage, angled to middle, and diagonal to top stage. Crenellated parapet with fine gargoyles.

GRADE: II

HERITAGE LIST NUMBER: 1149951

165 Great Ouseburn St Mary

LOCATION: Main Street, Great Ouseburn, Knaresborough

POSTCODE: YO26 9RG

OS GRID REF: SE449618

DATE: C12 and C13

DESCRIPTION: Norman tower with C13 belfry. 3 stages on chamfered plinth. Inserted lancet in first stage, west face, and slit lights to south on first and second stages: second-stage light has reset corbel-head above. Clock face to west in second stage. Bell openings on all faces are paired round-headed chamfered lights with squared centre shafts, closed by pierced shutters. Chamfered belfry string-course; embattled parapet with crocketed angle pinnacles above waterspouts to east, west and south. Pyramidal roof.

GRADE: II*

HERITAGE LIST NUMBER: 1150284

166 Grinton St Andrew

LOCATION: B6270, opposite The Bridge pub, Grinton

POSTCODE: DL11 6HH

OS GRID REF: SE046984

DATE: Late C12

DESCRIPTION: 3-storey embattled tower, 2 stair vents on south side, also bench mark; clock; cross window on all faces of belfry stage.

GRADE: I

HERITAGE LIST NUMBER: 1301991

167 Guiseley St Oswald

LOCATION: The Green, Guiseley

POSTCODE: LS20 9BB

OS GRID REF: SE194422

DATE: C14–C15

DESCRIPTION: Perpendicular 3-stage tower with diagonal buttresses has a chamfered 3-light west window with Perpendicular tracery, a clock face at the second stage, drip-mould to the third stage and cusped 2-light louvred belfry windows above this, a machicolated embattled parapet supported by plain gargoyles, with crocketed corner pinnacles; on south side a small 2-centred arched chamfered doorway.

GRADE: I

HERITAGE LIST NUMBER: 1135598

168 Halifax
St John the Baptist

LOCATION: Church Street, Halifax

POSTCODE: HX1 1QL

OS GRID REF: SE098252

DATE: 1449–82

DESCRIPTION: A south-west tower was intended, but was eventually built at the west corner instead: begun in 1449, it was finally completed in 1482.

GRADE: I

HERITAGE LIST NUMBER: 1133928

169 Hampsthwaite
St Thomas a Becket

LOCATION: Church Street, Hampsthwaite, Harrogate

POSTCODE: HG3 2HB

OS GRID REF: SE259591

DATE: C15

DESCRIPTION: Full-height stepped diagonal buttresses, fine west door in hollow-moulded Gothic arch with hood-mould, restored 3-light pointed Perpendicular window above. 3-light window with reticulated tracery to centre stage, south face; 3-light pointed Perpendicular belfry window to each side. All openings with hood-moulds. Eaves string-course, battlemented parapet with central triangular pilasters, and weathervane.

GRADE: II

HERITAGE LIST NUMBER: 1315374

170 Harewood All Saints

LOCATION: Harewood Park, Harewood

POSTCODE: LS17 9LG

OS GRID REF: SE314450

DATE: C15

DESCRIPTION: Low squat tower embraced by aisles with lean-to roofs has pointed-arch doorway of 3 orders with hood-mould and carved-face stops with above 5-light west window with panel tracery. Buttresses at junction with aisles. Chamfered band, recessed second stage has 2-light belfry window to each face. That to south has large clock face. Embattled parapet with corner rainwater chutes.

GRADE: I

HERITAGE LIST NUMBER: 1266157

171 Hauxwell St Oswald

LOCATION: Hauxwell, Leyburn

POSTCODE: DL8 5LP

OS GRID REF: SE165931

DATE: C13 with top storey c.1600

DESCRIPTION: 4-storey tower with offset angle buttresses. Ground floor 2 lancet windows; first and second floors 1 lancet. 2-light trefoil-headed belfry opening. Embattled parapet. West side: C19 window of 2 trefoil-headed lights with cinquefoil above; 2-light double-chamfered belfry opening. North side: belfry opening of 2 trefoil-headed lights with hollow chamfers. East side: belfry opening of 2 pointed lights in double-chamfered surround.

GRADE: I

HERITAGE LIST NUMBER: 1179762

172 Heptonstall St Thomas a Becket

LOCATION: Church Lane, Heptonstall

POSTCODE: HX7 7NT

OS GRID REF: SD986280

DATE: C13 and C15

DESCRIPTION: Church founded 1260 (fabric of this date survives in tower), raised mid-C15 when bells were inserted c.1440. Embattled tower has east stair-turret. Sheer wall with angle buttresses. West face has 2-over-3-light pointed-arch window with cusped heads and quatrefoil to apex. Hood-mould has beak-head and human mask stops. North and south faces have smaller 2-light traceried window with clear break in stonework over. This added stage has slightly flatter arched 2-light windows to each face with traceried heads to belfry. False machicolation to embattled parapet with chamfered moulded castellation of similar section to plinth. North and east faces have wide circle for clock of 1809 now in new church.

GRADE: II*

HERITAGE LIST NUMBER: 1226170

173 Hornby St Mary

LOCATION: Church View, Hornby, near Bedale

POSTCODE: DL8 1NH

OS GRID REF: SE222937

DATE: C11 and late C15

DESCRIPTION: West tower of 4 stages, the lower 3 of c.1080, the upper late C15. Round-arched windows to first and second stages, with clock on second stage; original belfry opening on third stage of 2 lights, divided by a recessed baluster with cushion capital, with round-arched lintels; double-chamfered 2-light belfry opening on fourth stage with mullion and transom, trefoiled segmental-pointed heads to lights and round-headed tracery above; C19 crenellated parapets supported on corbels; corner pinnacles recently removed. West side of tower has plinth, corner pilaster buttresses and doorway with slab lintel below round arch; north side of tower has round-arched light vent on first stage.

GRADE: I

HERITAGE LIST NUMBER: 1318321

174 Horton in Ribblesdale St Oswald

LOCATION: Main Street, Horton in Ribblesdale

POSTCODE: BD24 0HB

OS GRID REF: SD810721

DATE: C15

DESCRIPTION: West tower, segmental-pointed west entrance with moulded surround and hood-mould, plank door. 3-light Perpendicular window with trefoil heads, central light attached to mullion which pierces arch above, hood-mould. Small trefoil-headed window above and 2-light louvred window to bell stage with trefoil heads and hood-mould. Diagonal buttresses, string-course at base of embattled parapet.

GRADE: I

HERITAGE LIST NUMBER: 1132264

175 Hubberholme St Michael & All Angels

LOCATION: Stubbing Lane, Hubberholme

POSTCODE: BD23 5JE

OS GRID REF: SD925783

DATE: C12

DESCRIPTION: West tower, a narrow slit window to lower stage, south side, a rectangular window with leaded lights to the west face, a slit window above. Belfry windows of 1 and 2 lights with stone louvres; a drip-moulding between the stages and below the battlemented parapet which retains the line of a pitched roof on the east side.

GRADE: II*

HERITAGE LIST NUMBER: 1167280

176 Ilkley All Saints

LOCATION: Church Street, Ilkley

POSTCODE: LS29 9DS

OS GRID REF: SE116478

DATE: C14–C15

DESCRIPTION: The 3-stage Perpendicular tower has diagonal buttresses, embattled parapet and corner pinnacles. In the lower stage is a 3-light west window and stair-turret door in the south face. The second stage has a C19 2-light north window, and round clock on the south face. The upper stage has unrestored 2-light openings with louvres.

GRADE: II*

HERITAGE LIST NUMBER: 1133506

177 Ingleton
St Mary the Virgin

LOCATION: Main Street, Ingleton

POSTCODE: LA6 3EB

OS GRID REF: SD695732

DATE: C15

DESCRIPTION: West pointed entrance with deeply chamfered surround, hood-mould, plank door. Tall 3-light window with c.1887 mullions, hood-mould. 2-light louvred window with trefoil heads to bell stage. Diagonal buttresses, string-course to base of embattled parapet with 4 finials.

GRADE: II*

HERITAGE LIST NUMBER: 1252726

178 Kellington
St Edmund

LOCATION: Church Lane, Kellington village

POSTCODE: DN14 0NF

OS GRID REF: SE548246

DATE: C12–C15

DESCRIPTION: Lower stages of tower C12. Mid-C14 belfry stage and C15 stepped diagonal and angle buttresses. To first stage an inserted 3-light west window with Perpendicular tracery under hood-mould. First and second stage bands. Second stage blind. 2-light Perpendicular bell-openings to each side. Battlements.

GRADE: I

HERITAGE LIST NUMBER: 1148402

179 Kildwick St Andrew

LOCATION: Kirkgate, Kildwick

POSTCODE: BD20 9BB

OS GRID REF: SE012459

DATE: C14

DESCRIPTION: Tower of 2 stages with angle buttresses, slightly projecting stair-turret to south-east corner; west door in double-chamfered arch and 3-light Perpendicular-style window above, both restored 1860; small trefoil-headed niche below the belfry window which is of 2 lights with Perpendicular tracery to each face. Corbelled embattled parapet. Large clock face on south side with raised border inscription: "THE GIFT OF W. CURRER OF STEETON / LATE CITIZEN OF LONDON 1709 / RENEWED BY SUBSCRIPTION / REDEEM THE TIME / 1828." The ornate weathervane above the stair-turret has C and S scrolls, pennant and ball finial.

GRADE: I

HERITAGE LIST NUMBER: 1132175

180 Kippax St Mary

LOCATION: Church Lane, Kippax

POSTCODE: LS25 7HF

OS GRID REF: SE417304

DATE: C11 and C13

DESCRIPTION: Anglo-Saxon or early Norman tower (top stage remodelled probably in C13), has diagonal buttresses of that period, a round-headed lancet in each of the south and west sides of the first stage, 2-centred arched belfry windows of 2 cusped lights with stone louvres, and an embattled parapet with C19 crocketed corner pinnacles.

GRADE: I

HERITAGE LIST NUMBER: 1237399

181 Kirby Hill St Peter & St Felix

LOCATION: Village Green, Kirby Hill, Richmond

POSTCODE: DL11 7JJ

OS GRID REF: NZ140066

DATE: 1397

DESCRIPTION: 3-stage west tower, dated 1397: chamfered plinth; stepped diagonal buttress to left, with worn inscription "AD MCCCLXXXXVII", and below it a later inscription, "This Church built Anno Dom 1397 M.T.", and at the top of the buttress a carved figure with musical instrument; stair-turret to right with chamfered light vents; chamfered pointed-arch 1-light ground-floor window with label; clock and small trefoiled light with hood-mould on second stage, repeated on north side; offset third stage with flat-headed belfry opening of 2 trefoiled lights with hood-mould repeated on all other sides; string with gargoyle, and shields at corners, below crenellated parapet. West ground-floor tower window, hollow double-chamfered, of 2 trefoiled lights and Perpendicular tracery above; hood-mould with head stops. Against the south side is a tombstone commemorating Lucy, wife of Rev. James Stubbs, who was buried in the wall in 1805.

GRADE: I

HERITAGE LIST NUMBER: 1301472

182 Kirk Hammerton St John the Baptist

LOCATION: Church Street, Kirk Hammerton, Lower Nidderdale

POSTCODE: YO26 8DA

OS GRID REF: SE465555

DATE: C11

DESCRIPTION: Anglo-Saxon tower of 2 stages with 2-round-headed-light belfry windows with colonnette. Plain projecting band below window level. Shallow pyramidal roof.

GRADE: I

HERITAGE LIST NUMBER: 1149932

183 Kirkburton All Hallows

LOCATION: Shelley Lane, Kirkburton

POSTCODE: HD8 0SJ

OS GRID REF: SE198125

DATE: C15 and C19

DESCRIPTION: Perpendicular west tower, though upper part said to be C19. The tower has a fine Early English doorway, presumably reset, with heavily moulded arch and colonnette to each side, and large dog-tooth band around the whole doorway. Above it is a large 3-light window with Perpendicular tracery. 4-light mullioned-and-transomed bell-chamber openings with traceried heads. Crenellated parapet.

GRADE: I

HERITAGE LIST NUMBER: 1313318

184 Kirkby Fleetham St Mary

LOCATION: Kirkby Lane, Kirkby Fleetham

POSTCODE: DL7 0SU

OS GRID REF: SE281957

DATE: C15

DESCRIPTION: West tower of 3 stages, offset diagonal buttresses. To south a stair-tower with 2 small chamfered slit openings, band to belfry stage, 2-light Y-traceried pointed belfry openings. Moulded band, ashlar embattled parapet. Perpendicular west window of 3 lights with hood-mould.

GRADE: II*

HERITAGE LIST NUMBER: 1150928

185 Kirkby in Malhamdale St Michael the Archangel

LOCATION: Pig Lane, Kirkby in Malhamdale

POSTCODE: BD23 4BS

OS GRID REF: SD894610

DATE: C15

DESCRIPTION: 2-stage tower with diagonal buttresses. Ground floor west window of 1879 has 3 lights and hood-mould. South side has trefoil-headed niche above level of clerestory and panel inscribed "E:IR". Clock face on east and west sides. Belfry opening on each side has 3 lights, breaking into 4 trefoil-headed lights and continuous hood-mould. There are 5 shields at the south-west and south-east corners of the tower. Embattled parapet.

GRADE: I

HERITAGE LIST NUMBER: 1132389

186 Kirkby Malzeard St Andrew

LOCATION: Church Street, Kirkby Malzeard, Ripon

POSTCODE: HG4 3RT

OS GRID REF: SE235745

DATE: C15

DESCRIPTION: West tower of 3 stages, plinth with carved frieze, full-height offset diagonal buttresses and buttressed stair-tower rising to second stage. First stage: large Perpendicular-traceried 3-light pointed-arch west window with hood-mould. Second stage: a small 1-light ogee-headed opening to south. Band. Belfry windows, basket-arched, with 3 chamfered cusped lights and hood-mould, embattled parapet with pinnacles to corners.

GRADE: I

HERITAGE LIST NUMBER: 1173967

187 Kirkby Wharfe St John the Baptist

LOCATION: Green Lane, Kirkby Wharfe

POSTCODE: YO7 4ES

OS GRID REF: SE376848

DATE: C15

DESCRIPTION: West tower of 3 stages. Plinth and offset diagonal buttresses. Clock face below a C15 2-light pointed-arch cusped belfry window. Stone band, embattled parapet. West side has large 3-light Perpendicular window with hood-mould; above it a 1-light cusped opening and a belfry window similar to other sides.

GRADE: II*

HERITAGE LIST NUMBER: 1281456

188 Kirkheaton
St John the Baptist

LOCATION: Church Lane, Kirkheaton

POSTCODE: HD5 0BH

OS GRID REF: SE179172

DATE: C15

DESCRIPTION: The 3-tier tower has later south porch, diagonal buttresses with offsets to first tier, 2-light bell-chamber openings with tracery, and crenellated parapet with crocketed pinnacles. 3-light west window.

GRADE: II*

HERITAGE LIST NUMBER: 1135391

189 Kirklington
St Michael

LOCATION: Howgrave Road, Kirklington

POSTCODE: DL8 2NJ

OS GRID REF: SE318811

DATE: C15

DESCRIPTION: Perpendicular west tower of 4 tall stages. Offset diagonal buttresses and full-height stair-tower to south corner which rises above parapet with sloping roof. 2 middle stages have small rectangular chamfered openings to south. Belfry has flat-headed mullion and transom 2-light opening, with cusped heads to lights, to each side. Projecting embattled parapet. 3-light west window has pointed arch with hood-mould and rectilinear tracery. North side of tower has a clock face to third stage.

GRADE: I

HERITAGE LIST NUMBER: 1150766

190 Kirkthorpe, Warmfield St Peter

LOCATION: Kirkthorpe Lane, Kirkthorpe, Warmfield

POSTCODE: WF1 5SZ

OS GRID REF: SE362210

DATE: C14

DESCRIPTION: The 2-stage tower has diagonal buttresses, with offsets to west angles only, rising to just below second-stage band. C19 west window of 3 cusped lights with panel tracery. North-east stair-turret has lean-to roof and arrow slits. South face has inserted C19 window. 2-light belfry window to each face has Y-tracery and cusped lights. Set just below is small chamfered light. Corner gargoyles. Ashlar embattled parapet with corner crocketed pinnacles.

GRADE: II*

HERITAGE LIST NUMBER: 1313216

191 Leathley St Oswald

LOCATION: Leathley Lane, Leathley

POSTCODE: LS21 2LD

OS GRID REF: SE232470

DATE: Early C12

DESCRIPTION: Tower of 4 stages, slight quoins, a narrow single-splayed window to stages 2 and 3 and a larger, round-headed belfry window. Pyramidal roof.

GRADE: I

HERITAGE LIST NUMBER: 1174461

192 Little Ouseburn Holy Trinity

LOCATION: Church Lane, between Little and Great Ouseburn

POSTCODE: YO26 9SH

OS GRID REF: SE452611

DATE: C11 with C15 additions

DESCRIPTION: 3-stage C11 west tower with C15 embattled parapet and pinnacles. Inserted round-headed lights to south and west. Bell openings are paired round-headed louvred lights with centre shafts, above belfry string-course. Saddleback roof with waterspouts to north and south.

GRADE: I

HERITAGE LIST NUMBER: 1150296

193 Long Preston St Mary the Virgin

LOCATION: Church Street, Long Preston

POSTCODE: BD23 4NJ

OS GRID REF: SD837581

DATE: Late C14/early C15

DESCRIPTION: 3-stage tower, diagonal buttresses, two 1-light chamfered windows to first 2 stages. Bell stage has segmental-pointed 2-light windows with cinquefoil heads and quatrefoil above geometrical tracery. Clock face to west. Crocketed finials.

GRADE: I

HERITAGE LIST NUMBER: 1157818

194 Marton-in-Craven St Peter

LOCATION: Church Lane, Marton-in-Craven

POSTCODE: BD23 3LR

OS GRID REF: SD908507

DATE: C12

DESCRIPTION: Very plain rubble tower with low battlements. Very small round-headed windows near ground and small 2-light belfry openings.

GRADE: II*

HERITAGE LIST NUMBER: 1167226

195 Melsonby St James the Great

LOCATION: Church Row, Melsonby

POSTCODE: DL10 5NU

OS GRID REF: NZ201085

DATE: C13

DESCRIPTION: 3-stage tower. Moulded plinth and offsets. Ashlar pilaster buttresses at left corner and in centre, on lower second stage, with chamfered light vent on lowest stage of left buttress, and lancet window on second stage. 2-light belfry openings in shafted, pointed-arch surround. C19 corbelled battlemented parapet. West side of tower: moulded plinth and offsets, corner and central buttresses, with lancet windows on lowest and second stages of central buttress, with lozenge-shaped clock above on second stage, and belfry opening as before. North side of tower: as before, but with lancet window on second stage only and without clock. East side of tower: belfry opening as before, with chamfered hood-mould.

GRADE: II*

HERITAGE LIST NUMBER: 1167040

196 Methley St Oswald

LOCATION: Churchside, next to cricket field, Methley

POSTCODE: LS26 9BH

OS GRID REF: SE391266

DATE: C14 and C15

DESCRIPTION: 3-stage Perpendicular-style tower with diagonal buttresses has west door, 3-light window with Perpendicular tracery, louvred 2-light belfry windows, corbelled-out embattled parapet with crocketed corner pinnacles.

GRADE: I

HERITAGE LIST NUMBER: 1135664

197 Middleham St Mary & St Alkelda

LOCATION: Church Street, Middleham

POSTCODE: DL8 4PQ

OS GRID REF: SE126878

DATE: C15

DESCRIPTION: 3-storey embattled tower with stepped diagonal buttresses to west. 3-light Perpendicular west window. To south, single-chamfered doorway to stairs, 3-light vents above, small first-floor window with trefoil head. On all 4 sides, belfry openings of 2 lights with cinquefoil heads. Corner finials.

GRADE: I

HERITAGE LIST NUMBER: 1318544

198 Middleton Tyas St Michael & All Angels

LOCATION: Middleton Tyas, Richmond

POSTCODE: DL10 6RJ

OS GRID REF: NZ235055

DATE: C13

DESCRIPTION: Tower of 4 stages. C19 2-light west window and pointed second-stage light vent; south-east stair-turret with light vents; chamfered northern third-stage light vent; 2-light belfry openings; corbelled parapet.

GRADE: I

HERITAGE LIST NUMBER: 1131579

199 Normanton All Saints

LOCATION: High Street, Normanton

POSTCODE: WF6 1NR

OS GRID REF: SE387225

DATE: C15

DESCRIPTION: Perpendicular-style west tower of 3 stages. Diagonal buttresses with 4 offsets, moulded plinth, restored 3-light west window with cavetto-moulded reveal and hood-mould with small figured stops (worn), above this a small ogee-headed niche with a carved angel beneath and crocketed pinnacles above, a drip-mould to the slightly set-back belfry stage, with clock faces in the centre on all sides, belfry windows of 2 cusped lights with double-chamfered reveals and hood-moulds with carved stops, a corbel-table to an embattled parapet and crocketed corner pinnacles (restored); the south side has 3 chamfered lancets irregularly disposed at the second stage, and the south-west corner has very small loop-lights to the integral staircase.

GRADE: II*

HERITAGE LIST NUMBER: 1253747

200 Otley
All Saints

LOCATION: Kirkgate, Otley

POSTCODE: LS21 3HS

OS GRID REF: SE201453

DATE: C14 or C15

DESCRIPTION: Western tower with embattled parapet and small lead-covered wooden spire.

GRADE: I

HERITAGE LIST NUMBER: 1200290

201 Pannal
St Robert of Knaresborough

LOCATION: Main Street, Pannal, Knaresborough

POSTCODE: HG3 1JZ

OS GRID REF: SE306517

DATE: C15–C16

DESCRIPTION: The Perpendicular 3-stage tower has diagonal buttresses and embattled parapet. In the lower stage is a segmental-pointed west doorway with broad chamfer, and ribbed door. Above it is a 3-light window. The second stage has small square-headed windows beneath round clock faces, and square-headed 2-light bell stage openings have uncusped round-headed lights.

GRADE: II*

HERITAGE LIST NUMBER: 1149449

202 Penistone St John the Baptist

LOCATION: Shrewsbury Road, Penistone

POSTCODE: S36 6DY

OS GRID REF: SE246033

DATE: Late C14 or C15

DESCRIPTION: Tall Perpendicular, 3-stage tower with diagonal buttresses. Large west door with 2 orders of mouldings enriched with stars and masks. 3-light traceried window above. Transomed and traceried bell-chamber openings. Gargoyles, castellated parapets and crocketed pinnacles.

GRADE: I

HERITAGE LIST NUMBER: 1314709

203 Pickhill All Saints

LOCATION: Money Hill, Pickhill

POSTCODE: YO7 4JP

OS GRID REF: SE347837

DATE: Late C15

DESCRIPTION: 3-stage tower with offset diagonal buttresses and wide angle buttress to east of south front which forms stair-tower with 3 slit openings; 1-light chamfered opening and clock face to second stage; 2-light flat-headed cusped belfry window; band; embattled parapet with pinnacles at each corner. West window, C19, 3-light pointed arch with hood-mould.

GRADE: II*

HERITAGE LIST NUMBER: 1204105

204 Richmond
St Mary the Virgin

LOCATION: Station Road, Richmond

POSTCODE: DL10 7AQ

OS GRID REF: NZ174011

DATE: C15

DESCRIPTION: The Perpendicular 3-stage tower has diagonal buttresses with gabled offsets, and embattled parapet. In the lower stage is a 4-light west window. The second stage has a small narrow west window and clock on the north face. 2-light belfry openings have transoms and louvres.

GRADE: II*

HERITAGE LIST NUMBER: 1289814

205 Ripley
All Saints

LOCATION: Hollybank Lane, centre of village, Ripley

POSTCODE: HG3 3AY

OS GRID REF: SE283605

DATE: C15 and C16

DESCRIPTION: The upper stage of the tower is narrower, indicating the C16 heightening, and contains 2-light belfry openings with chamfered mullions; the lower stage west window is of 3 trefoil-headed lights. Crenellated parapet.

GRADE: II*

HERITAGE LIST NUMBER: 1174051

206 Romaldkirk St Romald

LOCATION: Middle Green, Romaldkirk, Barnard Castle

POSTCODE: DL12 9EE

OS GRID REF: NY995221

DATE: C15

DESCRIPTION: 3-stage tower has stepped diagonal buttresses; 3-light Perpendicular west window to first stage; scattered slit openings above; 2-light mullioned-and-transomed bell-openings with cinquefoil heads; embattled parapet with worn gargoyles and later corner pinnacles.

GRADE: I

HERITAGE LIST NUMBER: 1121857

207 Royston St John the Baptist

LOCATION: Church Street, Royston

POSTCODE: S71 4QZ

OS GRID REF: SE364112

DATE: C15

DESCRIPTION: Perpendicular style. Square 4-stage tower with diagonal buttresses. Moulded west doorway with 3-light window above. The third stage has a 5-sided oriel window with cusped lights which have shields below, and a pyramidal roof. Louvred, transomed belfry openings. Gargoyles, crenellated parapet and crocketed pinnacles.

GRADE: I

HERITAGE LIST NUMBER: 1151127

208 Scruton St Radegund

LOCATION: The Green, Scruton

POSTCODE: DL7 0QY

OS GRID REF: SE300926

DATE: C15

DESCRIPTION: West tower of 3 stages. To west, offset diagonal buttresses up to belfry. Clock to second stage. Chamfered band. Paired chamfered ogee-arched lights to belfry openings. Moulded string-course with gargoyle at each corner. Embattled parapets, pyramidal roof with weathervane.

GRADE: II*

HERITAGE LIST NUMBER: 1150897

209 Silkstone All Saints

LOCATION: High Street, Silkstone

POSTCODE: S75 4JQ

OS GRID REF: SE291058

DATE: C15

DESCRIPTION: Tall Perpendicular-style 2-stage tower with diagonal buttresses. Moulded west doorway with deeply set, hollow-chamfered traceried window above. 2-light transomed, traceried bell-chamber openings. Crenellated parapets with corner gargoyles and pinnacles.

GRADE: I

HERITAGE LIST NUMBER: 1151740

210 Skipton Holy Trinity

LOCATION: High Street, Skipton

POSTCODE: BD23 1NJ

OS GRID REF: SD990520

DATE: C15

DESCRIPTION: Tall Perpendicular-style west tower. Thin diagonal buttresses. Battlements and pinnacles.

GRADE: I

HERITAGE LIST NUMBER: 1131877

211 Slaidburn St Andrew

LOCATION: Church Street, Slaidburn

POSTCODE: BB7 3ER

OS GRID REF: SD710521

DATE: C15

DESCRIPTION: The 5-stage tower has angle buttresses and a solid parapet. The west door has a pointed head and jambs of 2 hollow-chamfered orders. The west window is of 3 cusped lights with panel tracery. Above, one over the other, are 2 niches, now without statues, having projecting decorated heads. Between them is a 1-light chamfered window. The bell-openings, probably early C15, have pointed heads and hoods, each having 2 cusped lights.

GRADE: I

HERITAGE LIST NUMBER: 1163738

212 South Cowton St Mary

LOCATION: Close to castle, Scorton Road

POSTCODE: DL7 0JB

OS GRID REF: NZ293026

DATE: C15

DESCRIPTION: Perpendicular tower; flat-headed windows; 2-light ogee-arched belfry opening with flat hood-mould. Stair-turret with parapet. Embattled parapet with pinnacles.

GRADE: I

HERITAGE LIST NUMBER: 1294728

213 South Kirkby All Saints

LOCATION: Barnsley Road, South Kirkby

POSTCODE: WF9 3NT

OS GRID REF: SE453110

DATE: C15

DESCRIPTION: Perpendicular 4-stage tower has diagonal buttresses with crocketed finials on the offsets, weathered plinth, and drip-bands on 4 levels; restored west doorway has set-in shafts and 2 orders of moulding, nail-head ornament on the head, hollow-moulded hood-mould with angel stops (worn) and ogee cresting; 4-centred arched west window (boarded at time of survey, 1987), small loop-light to third stage, tall transomed 3-light belfry windows with restored tracery and hood-moulds with figured stops, gargoyles in the band above, and embattled parapet with crocketed pinnacles at the corners and in the centre of each side; clock faces in north and south sides of third stage.

GRADE: I

HERITAGE LIST NUMBER: 1227519

214 Spennithorne St Michael

LOCATION: Main Street, Spennithorne, Leyburn

POSTCODE: DL8 5PR

OS GRID REF: SE137889

DATE: C14

DESCRIPTION: C14 tower with offset angle buttresses and projecting stair-turret to south with 6-light vents and bench mark to west face of projection. Ground floor: 2-light west window. First floor: clock on west face, 1-light window with trefoil head under flat hood to north and south. Second floor: belfry openings of 2 lights, with trefoil heads and blind quatrefoil in plate tracery above, on all 4 sides. Embattled parapet corbelled out on heads, with gargoyles of goblin holding spout, weathervane on north-west corner.

GRADE: I

HERITAGE LIST NUMBER: 1179485

215 Spofforth All Saints

LOCATION: Church Lane, Spofforth

POSTCODE: HG3 1AF

OS GRID REF: SE364510

DATE: C15

DESCRIPTION: The west tower has 4 stages, Perpendicular windows, offset diagonal buttresses, clock and battlemented parapet.

GRADE: II*

HERITAGE LIST NUMBER: 1149981

216 Stanwick St John

LOCATION: Stanwick

POSTCODE: DL11 7RS

OS GRID REF: NZ185119

DATE: C13

DESCRIPTION: Tower of 3 stages, with chamfered offsets; quoins; stepped diagonal buttresses; 1-light ringing-chamber window with shouldered head and 2-light shafted belfry opening, spout from roof, battlemented parapet; to west, same openings plus ground-floor lancet window above central buttress, and on ground between central and south-west corner buttress a medieval stone coffin on its side; to north, 5-sided C19 stair-turret with shoulder-headed doorway and quatrefoil light vents, and belfry opening as before; to east, chamfered doorway with hood-mould giving onto nave roof, steep nave roofline, and belfry opening as before.

GRADE: I

HERITAGE LIST NUMBER: 1167219

217 Thorner St Peter

LOCATION: Church View, end of Main Street, Thorner

POSTCODE: LS14 3ED

OS GRID REF: SE380405

DATE: C15

DESCRIPTION: Perpendicular-style, 2-stage embattled tower: plinth, angle buttresses with offsets. 3-light west window with octagonal clock face above; 2-light belfry windows. South face has small arched doorway to base and 2-light window with chamfered light above. Parapet carried on corbel-table, corner pinnacle.

GRADE: II

HERITAGE LIST NUMBER: 1227581

218 Thornhill St Michael & All Angels

LOCATION: Church Lane, Thornhill

POSTCODE: WF12 0JZ

OS GRID REF: SE254189

DATE: C15

DESCRIPTION: Perpendicular west tower, probably C15, though the base may be earlier. The square tower is in 2 tall stages with reducing diagonal buttresses. West door with moulded surround, west window probably C19. 8-light mullioned-and-transomed bell-chamber openings with 5-foiled lights and traceried heads, crenellated parapet on corbel brackets and 8 crocketed pinnacles.

GRADE: I

HERITAGE LIST NUMBER: 1200754

219 Thornton-in-Craven St Mary

LOCATION: Church Road, Thornton-in-Craven

POSTCODE: BD23 3TS

OS GRID REF: SD901483

DATE: Early C16

DESCRIPTION: The early C16 tower has a 3-light window at ground level and 3-light louvred belfry windows. Diagonal buttresses.

GRADE: I

HERITAGE LIST NUMBER: 1167634

220 Thornton Watlass St Mary

LOCATION: Thornton Watlass, Ripon

POSTCODE: HG4 4AJ

OS GRID REF: SE233854

DATE: C15

DESCRIPTION: Perpendicular-style tower of 4 stages. Offset clasping buttresses with stair-tower in that to south-east having chamfered lights. Second stage has trefoil-headed 1-light window. Third stage has 1-light pointed-arch window, belfry has three 1-light trefoil-headed windows set in a double-chamfered surround with flat lintel above each light, corbelled cornice with embattled parapet, pinnacles to each corner.

GRADE: II

HERITAGE LIST NUMBER: 1314992

221 Well St Michael

LOCATION: Church Street, Well, Bedale

POSTCODE: DL8 2QA

OS GRID REF: SE268820

DATE: C12 and C15

DESCRIPTION: West tower: C12 to tall first stage, Perpendicular to upper 3 stages. First stage has 2 small chamfered 1-light openings to south side. Wide band to second stage which is blank. Wide band to third stage and paired cusped lancets, blocked to base. Belfry has 2 large shallow-arched openings next to each other, each contains a 2-light opening with cusped pointed-arch heads and a transom. At this level the corners are cut back to form a small diagonal buttress. Band, embattled parapet. West window is set in segmental-headed opening but window is flat-headed with 2 cusped lights. Clock faces to west and east below belfry.

GRADE: I

HERITAGE LIST NUMBER: 1315190

222 West Tanfield St Nicholas

LOCATION: Main Street, West Tanfield

POSTCODE: HG4 5JQ

OS GRID REF: SE268787

DATE: C15

DESCRIPTION: Perpendicular west tower of 3 stages. Offset diagonal buttresses up to belfry level. To south side a buttress-like stair-tower up to the belfry, to bottom is a segmental-arched doorway, above very small 1-light chamfered openings. Stone bands between stages. Blank second stage. Belfry openings have pointed-arch head with hood-moulds, cusped tracery of 2 lights. Projecting embattled parapet. Window has pointed-arch head with hood-mould, of 3 lights with rectilinear tracery. Clock face below belfry to west facade.

GRADE: I

HERITAGE LIST NUMBER: 1150775

223 West Witton St Bartholomew

LOCATION: Church Street, behind Main Street, West Witton

POSTCODE: DL8 4LU

OS GRID REF: SE061885

DATE: C16

DESCRIPTION: Tower of 3 storeys. Quoins. West ground-floor window of 2 uncusped pointed lights below hood-mould with head stops. Light vent to south side of ringing chamber. Clock to west and south faces. On all sides, belfry openings of 2 trefoil-headed lights below hood-mould. String. Gargoyle to south. Battlemented parapet with corner pinnacles.

GRADE: II

HERITAGE LIST NUMBER: 1318560

224 Whixley
The Ascension

LOCATION: Church Street, Whixley, Kirkby on the Moor

POSTCODE: YO26 8AR

OS GRID REF: SE442584

DATE: C14–C15

DESCRIPTION: West tower of 3 stages with Perpendicular window to belfry, embattled parapet with crocketed finials and pyramidal roof.

GRADE: II*

HERITAGE LIST NUMBER: 1189906

225 Woodkirk
St Mary

LOCATION: Dewsbury Road, Woodkirk

POSTCODE: WF12 7JL

OS GRID REF: SE272250

DATE: Late C12

DESCRIPTION: 3-stage tower, embattled c.1911, has battered plinth, roll-moulded strings. West face has lancet window to first stage, clock to second stage and 2-light plate tracery belfry windows recessed inside elliptical arches and having colonnettes with cushion-headed capitals. Battlements have corner pinnacles. Weathervane. Earlier steeply pitched roofline visible on east face.

GRADE: II

HERITAGE LIST NUMBER: 1250737

226 Woolley
St Peter

LOCATION: Church Street, Woolley

POSTCODE: WF4 2JU

OS GRID REF: SE320131

DATE: C15

DESCRIPTION: The 2-stage embattled tower has diagonal buttresses with offsets. Chamfered plinth, moulded band. West doorway with double-channelled surround has hood-mould. 3-light west window above with cusped lights and panel tracery; on south face 4 arrow slits one above the other to light stair. 2-light belfry openings have cusped lights with quatrefoil over to each face; clock face set under to north. Projecting rainwater spouts at base of parapet with corner and central crocketed pinnacles.

GRADE: I

HERITAGE LIST NUMBER: 1200723

227 Wragby
St Michael & Our Lady

LOCATION: Nostell Priory, Wragby

POSTCODE: WF4 1RA

OS GRID REF: SE408173

DATE: C15–early C16

DESCRIPTION: Perpendicular-style unbuttressed tower with a chamfered plinth; 2-centred arched west doorway with 2 orders of hollow moulding, and a hood-mould; a 3-light west window with Perpendicular tracery and a hood-mould, and above this a niche with cusped ogee head and a finial flanked with 2 carved wheel patterns on each side, a drip-mould to the belfry stage. 2-light louvred belfry windows with Perpendicular tracery and hood-moulds, and an embattled parapet with crocketed corner pinnacles.

GRADE: I

HERITAGE LIST NUMBER: 1253565

Diocese of
Liverpool

228 Farnworth St Luke

LOCATION: Pitt Lane, Farnworth

POSTCODE: WA8 9HS

OS GRID REF: SJ514877

DATE: C14

DESCRIPTION: Crenellated west tower has corner buttresses, curvilinear west window and smaller louvred windows to all faces at bell stage.

GRADE: II*

HERITAGE LIST NUMBER: 1130417

229 Hale St Mary

LOCATION: Church Road, Hale

POSTCODE: L24 4BZ

OS GRID REF: SJ471820

DATE: C14

DESCRIPTION: Square west tower with corner buttresses and crenellations. Boarded oak door in Gothic opening with drip-mould and simple geometrical window above. Also curvilinear louvre windows to all faces at bell stage.

GRADE: II

HERITAGE LIST NUMBER: 1130454

230 Huyton St Michael

LOCATION: Blue Bell Lane, Huyton

POSTCODE: L36 9XJ

OS GRID REF: SJ442911

DATE: C14

DESCRIPTION: Tower has moulded base; diagonal buttresses with pinnacles at bell stage. West entrance has continuous mouldings; frieze over has roundels and quatrefoils with shields; drip-mould over; bell stage has sill course and 3-light louvred bell-openings with uncusped Perpendicular tracery; cornice with gargoyles and embattled parapet with obelisk pinnacles with iron vanes.

GRADE: II*

HERITAGE LIST NUMBER: 1075535

231 Ormskirk
St Peter & St Paul

LOCATION: Park Road, Ormskirk

POSTCODE: L39 3AP

OS GRID REF: SD413084

DATE: c.1540–50

DESCRIPTION: The west tower has a massive square plan, and is of 3 unequal stages (the first forming the lower half); with diagonal buttresses, a weathered plinth, weathered coping to the first stage, set-back upper stages with weathered sill band, and an embattled parapet with crocketed pinnacles in the centre and at the corners. It has a 4-centred arched west doorway with 3 orders of roll-moulding, a restored Perpendicular 3-light window above the doorway, a square-headed loop-light to the second stage, and coupled 2-centred arched 3-light belfry windows with intersecting tracery and stone louvres; and in each of the other sides a 3-light window offset eastward.

GRADE: II*

HERITAGE LIST NUMBER: 1221160

232 Upholland
St Thomas the Martyr

LOCATION: Church Street, Upholland

POSTCODE: WN8 0LW

OS GRID REF: SD523051

DATE: Late C15

DESCRIPTION: The tower, of 4 unequal stages, with diagonal buttresses which have 8 set-offs, has a low 2-centred arched west doorway with 3 orders of hollow moulding containing masks, shields, figures and Tudor roses, a large C19 2-centred arched 3-light west window with reticulated tracery, a drip-band, 2-light belfry windows with stone louvres, a short top stage with clock faces, a drip-band over this with grotesques, and an embattled parapet with corner pinnacles.

GRADE: I

HERITAGE LIST NUMBER: 1201666

Diocese of
Manchester

233 Deane
St Mary

LOCATION: 13–14 Junction Road, Deane

POSTCODE: BL3 4LT

OS GRID REF: SD693081

DATE: C14

DESCRIPTION: Undivided west tower, with ogee-arched west doorway and 3-light traceried window above. Paired bell-chamber lights, and clock in moulded stone surround on the south face. Embattled parapet with angle pinnacles.

GRADE: II*

HERITAGE LIST NUMBER: 1388085

234 Eccles
St Mary the Virgin

LOCATION: Church Street, Eccles

POSTCODE: M30 0DF

OS GRID REF: SJ779986

DATE: C15 with C13 base

DESCRIPTION: The castellated 3-stage tower has weathered diagonal buttresses, a west door, clock faces, a 3-light west window, 3-light belfry openings and crocketed corner pinnacles.

GRADE: I

HERITAGE LIST NUMBER: 1067498

235 Middleton St Leonard

LOCATION: St Leonard's Square, off New Lane, Middleton

POSTCODE: M24 6DE

OS GRID REF: SD872063

DATE: 1412 with top stage added c.1667

DESCRIPTION: The 4-stage west tower built in 1412 by Cardinal Thomas Langley, Bishop of Durham, has diagonal buttresses, a weathered plinth, 2-light windows on the first and second stages, clock faces on the third, and a curious weatherboarded top stage which was designed to accommodate a new peal of bells and has gables on each side.

GRADE: I

HERITAGE LIST NUMBER: 1162332

236 Prestwich St Mary the Virgin

LOCATION: Church Lane, Prestwich

POSTCODE: M25 1AN

OS GRID REF: SD811037

DATE: C15

DESCRIPTION: 3-stage battlemented tower.

GRADE: I

HERITAGE LIST NUMBER: 1067252

Diocese of

Newcastle

237 Alnwick St Michael

LOCATION: Baliffgate, Alnwick

POSTCODE: NE66 1ND

OS GRID REF: NU184137

DATE: C15

DESCRIPTION: The embattled tower has angle buttresses with 12 set-offs ending in pinnacles; the ground floor has a segmental-headed window with Perpendicular tracery, the second stage has paired narrow openings, and the third stage has a Perpendicular slatted window.

GRADE: I

HERITAGE LIST NUMBER: 1041546

238 Ancroft St Anne

LOCATION: Ancroft

POSTCODE: TD15 2TE

OS GRID REF: NU002451

DATE: C13

DESCRIPTION: West end of C12 nave was raised into a tower of C13. On the south side the tower has 2 lancet windows and 2 slit windows to the newel stair. On the west side the roof mark of the gable of the original towerless church can be seen, as can the band where the corbel-table was cut away. This side has 2 faucets and a C12 window with shafts. The corbel-table formerly continued on the north side. The tower was built to be defensible. It has a tunnel-vaulted ground floor blocking the C12 doorway, and a stone newel stair in the corner.

GRADE: II*

HERITAGE LIST NUMBER: 1042234

239 Bamburgh St Aidan

LOCATION: Church Street/Radcliffe Road, Bamburgh

POSTCODE: NE69 7AB

OS GRID REF: NU178350

DATE: C13 or early C14

DESCRIPTION: 4-stage tower with offsets. Y-tracery window on ground floor, lancet bell-openings; top storey C19 with battlements.

GRADE: I

HERITAGE LIST NUMBER: 1042269

240 Bolam St Andrew

LOCATION: Bolam Lake Country Park, Whalton

POSTCODE: NE61 3UA

OS GRID REF: NU092826

DATE: Late Saxon

DESCRIPTION: Ground floor has C18 lancets on south and west sides, with heads of original windows above (cf. interior). Small original windows on first floor and 2-light openings with baluster mullions on second floor. C12 window above and renewed parapet.

GRADE: I

HERITAGE LIST NUMBER: 1304102

241 Bywell
St Andrew

LOCATION: Bywell village, south of B6309

POSTCODE: NE43 7AE

OS GRID REF: NZ048615

DATE: Early C11

DESCRIPTION: Upper part of tower early C11, lower part of tower probably earlier. Tall tower with string-courses below and above belfry; apparent buttresses are remains of west gable of taller Saxon nave. Lower stage has original round-headed window on south, and similar window above C19 lancet on west; on south, just below belfry, high-level doorway with monolithic arched head and projecting imposts and bases which link to strip-work surround. 2-light belfry openings with baluster mullions, impost blocks and strip-work surrounds; circular sound holes in tympana and to either side.

GRADE: I

HERITAGE LIST NUMBER: 1044831

242 Bywell
St Peter

LOCATION: Bywell village, south of B6309

POSTCODE: NE43 7AD

OS GRID REF: NZ049613

DATE: Early C14

DESCRIPTION: Squat tower with chamfered set-back at mid-height on north and west. West wall shows doorway with chamfered arch and hood; to right blocked upper door with segmental relieving arch. Belfry openings are paired lancets except for crude square-headed east opening. Embattled parapet on hollow-chamfered string.

GRADE: I

HERITAGE LIST NUMBER: 1044828

243 Corbridge St Andrew

LOCATION: Market Place, Corbridge

POSTCODE: NE45 5TA

OS GRID REF: NY988644

DATE: Early C11

DESCRIPTION: Nave late C7 or C8, with 2-storey west porch raised into tower in early C11. The tower shows massive quoins; C19 3-light window set in original arched doorway, with worn saltire crosses on voussoirs of relieving arch, and round-headed window above. Similar window above nave roof on east is the original west window of taller Saxon nave, now incorporated in tower wall. Band below altered bell-openings with later imposts and pointed heads on old jambs. String-course above; tall crenellated parapet with blocked openings below on east and west.

GRADE: I

HERITAGE LIST NUMBER: 1303508

244 Edlingham St John the Baptist

LOCATION: Beside Edlingham Castle, off the B6341

POSTCODE: NE66 2BN

OS GRID REF: NU114091

DATE: c.1300

DESCRIPTION: 3-stage west tower has stepped chamfered plinth, band above plinth, and chamfered set-back below squat belfry. Lower stage has small rebated lancet on south only, the upper stages slatted chamfered loops except on east, which shows weathering of old high-pitched roof. Low pyramidal roof.

GRADE: I

HERITAGE LIST NUMBER: 1041990

245 Embleton Holy Trinity

LOCATION: To the west of Embleton village

POSTCODE: NE66 3BD

OS GRID REF: NU230225

DATE: Late C11–early C14

DESCRIPTION: 3-stage tower with chamfered set-backs between stages and below parapet. Lower part of tower late C11 or early C12 and heightened in early C14. Restored or C19 stepped buttresses flank C19 2-light west window. South wall of lower stage shows part-blocked trefoil-headed window above aisle roof; second stage has 2 square-headed windows on west; belfry has transomed openings of 2 trefoil-headed lights with quatrefoil spandrel; parapet with trefoil-headed open panels and 8 small pinnacles.

GRADE: I

HERITAGE LIST NUMBER: 1041822

246 Hartburn
St Andrew

LOCATION: B6343, Hartburn

POSTCODE: NE61 4JB

OS GRID REF: NZ090860

DATE: c.1200

DESCRIPTION: 3-stage tower has moulded plinth and strings; stepped set-back buttresses. C15 3-light west window with pierced panel tracery; wide lancets to second stage, blocked except on south; paired-lancet belfry openings with octagonal column mullions. Parapet on stepped corbels.

GRADE: I

HERITAGE LIST NUMBER: 1371047

247 Ingram
St Michael & All Angels

LOCATION: Beside the River Breamish in Ingram

POSTCODE: NE66 4LU

OS GRID REF: NU019163

DATE: C11 and C13

DESCRIPTION: Unbuttressed broad, square, west tower. Small C12 window in lower stage, paired lancets in bell stage. Plain parapet.

GRADE: II*

HERITAGE LIST NUMBER: 1276811

248 Kirkwhelpington St Bartholomew

LOCATION: Centre of Kirkwhelpington village

POSTCODE: NE19 2RS

OS GRID REF: NY996844

DATE: C12–C16

DESCRIPTION: West tower of several builds. Lower masonry C12 or C13. Low, late Perpendicular west door with Tudor lintel. Buttresses of various dates C13–C16. Battlements and crocketed pinnacles Perpendicular features. Victorian 3-light bell-openings with Perpendicular tracery features.

GRADE: II*

HERITAGE LIST NUMBER: 1044915

249 Longhoughton St Peter & St Paul

LOCATION: North End, Longhoughton

POSTCODE: NE66 3AG

OS GRID REF: NU243151

DATE: Early C12

DESCRIPTION: Broad squat west tower in 2 stages divided by chamfered set-back. Later medieval stepped buttresses on south and west and narrow round-headed lights in each wall. C19 2-light belfry openings in Saxon style, except on east where there is an old blocked opening and weatherings of 2 levels of nave roof; string below C19 embattled parapet.

GRADE: I

HERITAGE LIST NUMBER: 1303950

250 Morpeth St Mary the Virgin

LOCATION: St Mary's Field, Morpeth

POSTCODE: NE61 2QF

OS GRID REF: NZ197851

DATE: C14

DESCRIPTION: Largely C14, 2-storey tower with angle buttresses and 2-light bell-openings; low, lead C18 concave spire.

GRADE: I

HERITAGE LIST NUMBER: 1042763

251 Newburn St Michael & All Angels

LOCATION: Church Bank, Newburn

POSTCODE: NE15 8LT

OS GRID REF: NZ167653

DATE: C11

DESCRIPTION: 3-stage tower has blocked round-headed west door with inserted small round-headed window; similar window above; second stage has slit opening; belfry openings round-headed containing 2 round-arched lights and central shaft. Clasping buttresses; pyramidal roof and weathervane.

GRADE: I

HERITAGE LIST NUMBER: 1186036

252 Newcastle St John the Baptist

LOCATION: Grainger Street, Newcastle upon Tyne

POSTCODE: NE1 5JG

OS GRID REF: NZ247640

DATE: C14 and C15

DESCRIPTION: 3-stage tower with 3-light window, small window above, and bar tracery to 2-light belfry openings under battlemented parapet with corner pinnacles; octagonal clasping buttresses.

GRADE: I

HERITAGE LIST NUMBER: 1024728

253 Ovingham St Mary the Virgin

LOCATION: Main Road, Ovingham

POSTCODE: NE42 6BW

OS GRID REF: NZ085637

DATE: C11

DESCRIPTION: Unbuttressed west tower has 2-light round-arched bell-openings, with baluster mullion and pierced roundel over, in round-headed thin raised frame. Further small round-headed windows below, 2 on south side and one on west.

GRADE: I

HERITAGE LIST NUMBER: 1044940

254 Ponteland St Mary the Virgin

LOCATION: North Road, Ponteland

POSTCODE: NE20 9NU

OS GRID REF: NZ166729

DATE: C12

DESCRIPTION: West door with one order of columns and chevron moulding in the arch. 2 small original windows above. Several other C12 windows on returns, also several C14 cusped windows. Bell stage added C14.

GRADE: I

HERITAGE LIST NUMBER: 1370736

255 Stamfordham St Mary the Virgin

LOCATION: The Green, behind the Bay Horse Inn, Stamfordham

POSTCODE: NE18 0QQ

OS GRID REF: NZ076720

DATE: C13

DESCRIPTION: Tower of 3 storeys. Small lancets on ground and first floor; paired-lancet bell-opening with roll-moulded surround and central column with water-leaf capital.

GRADE: I

HERITAGE LIST NUMBER: 1370769

256 Warden
St Michael & All Angels

LOCATION: Centre of Warden village

POSTCODE: NE46 4SW

OS GRID REF: NY914664

DATE: Saxon

DESCRIPTION: Tall west tower of coursed sandstone with irregular large quoins. Small round-headed windows differing in detail, stepped-in C18 belfry stage with single broad pointed opening in each face, low pyramidal cap.

GRADE: I

HERITAGE LIST NUMBER: 1044988

257 Whalton
St Mary Magdalene

LOCATION: South edge of Whalton

POSTCODE: NE61 3XH

OS GRID REF: NZ131812

DATE: C11

DESCRIPTION: West tower probably C11. Tower on stepped and chamfered plinth, with later stepped buttress on south. Gothic west window with imposts and hood-mould, holding 16-pane sash with intersecting glazing bars. Single segmental-headed light on west below belfry, and single-handed 1796 clock face on east. Broad trefoiled bell-openings except for round-headed light on west. Embattled parapet.

GRADE: I

HERITAGE LIST NUMBER: 1247805

Diocese of

Sheffield

258 Adlingfleet All Saints

LOCATION: Church Lane, Adlingfleet

POSTCODE: DN14 8HS

OS GRID REF: SE844210

DATE: C15

DESCRIPTION: 2-stage tower: full-height diagonal buttresses with offsets, moulded plinth, slit lights to staircase in south-west angle. Tall first stage: large C15 pointed 4-light west window with Perpendicular tracery, hood-mould; single slit lights above to west, north and south sides. Stepped-in upper stage has 4-centred arched 2-light belfry openings with cinquefoiled lights, incised spandrels and hood-moulds. Moulded string-course, gargoyles to corners and centres of each side. Coped embattled parapet.

GRADE: I

HERITAGE LIST NUMBER: 1083144

259 Adwick-le-Street St Laurence

LOCATION: Village Street, Adwick-le-Street

POSTCODE: DN6 7AD

OS GRID REF: SE541086

DATE: Late C14–C15

DESCRIPTION: Perpendicular tower; chamfered plinth and moulded band, diagonal buttresses to lower stage. West side has pointed 3-light window in chamfered surround beneath clock; string-course with image niche; cusped iron clock face on south side; louvred, cusped 2-light belfry openings with pointed arches beneath string-course with gargoyles; embattled parapet with crocketed pinnacles (1 missing).

GRADE: II*

HERITAGE LIST NUMBER: 1151473

260 Aston
All Saints

LOCATION: Church Lane, Aston cum Aughton

POSTCODE: S26 2EE

OS GRID REF: SK468852

DATE: C14

DESCRIPTION: 3-stage tower. Chamfered plinth and moulded band. Diagonal offset buttresses to lower 2 stages. 3-light west window with restored Perpendicular tracery beneath pointed arch with hood-mould; 2 string-courses and offsets beneath 3-light belfry windows with ashlar aprons, louvres, panel tracery and pointed arches with hood-moulds returning as a string-course; crenellated parapet with crocketed pinnacles.

GRADE: I

HERITAGE LIST NUMBER: 1151917

261 Barnby Dun
St Peter & St Paul

LOCATION: Church Road, Barnby Dun

POSTCODE: DN3 1EB

OS GRID REF: SE614097

DATE: C14 and C15

DESCRIPTION: Perpendicular tower of 3 stages. Chamfered plinth, moulded band, full-height angle buttresses. String-course beneath 3-light west window having hollow-chamfered surround and hood-mould with head-carved stops; string-courses below and above second stage; 2-light belfry windows with louvres and hood-moulds with head-carved stops; string-course with north and south gargoyles beneath embattled parapet with 8 crocketed pinnacles.

GRADE: I

HERITAGE LIST NUMBER: 1151488

262 Bradfield St Nicholas

LOCATION: Town Gate, High Bradfield

POSTCODE: S6 6LG

OS GRID REF: SK267925

DATE: C14

DESCRIPTION: 3-stage west tower; plinth and tall diagonal buttresses. West door beneath 2-light cusped Y-tracery window. String-course beneath similar 2-light bell-openings. Another string-course beneath parapet with corner pinnacles. Clock on south face dated 1867.

GRADE: I

HERITAGE LIST NUMBER: 1192617

263 Braithwell St James

LOCATION: Micklebring Lane, Braithwell

POSTCODE: S66 7AS

OS GRID REF: SK530947

DATE: C15

DESCRIPTION: 3-stage Perpendicular ashlar tower. Chamfered plinth, moulded band and diagonal west buttresses; 3-light west window with head stops to hood-mould; string-courses below and above plain second stage; apron panels to louvred 3-light belfry windows; string-course beneath embattled parapet having 8 crocketed pinnacles.

GRADE: II*

HERITAGE LIST NUMBER: 1286514

264 Brodsworth St Michael & All Angels

LOCATION: In the grounds of Brodsworth Hall

POSTCODE: DN5 7XH

OS GRID REF: SE507072

DATE: c.1200

DESCRIPTION: Plinth, angle buttresses. C19 west window of 2 cusped lights. 2 offset bands, the upper partly decorated with saltire crosses. Later belfry stage with 2-light openings beneath chamfered, 4-centred arches. East belfry window has carved spandrels and round arch with clock above. Cavetto-moulded course with gargoyles beneath embattled parapet.

GRADE: II*

HERITAGE LIST NUMBER: 1191555

265 Burghwallis St Helen

LOCATION: Grange Lane, Burghwallis

POSTCODE: DN6 9JL

OS GRID REF: SE537120

DATE: C13

DESCRIPTION: 3-stage tower; chamfered plinth, large quoins to irregular rubble, offsets at string-courses between stages. Round-headed loops to west and south; small circular window reset in second stage on north side; belfry stage has paired-lancet openings divided by shafts; C15 top with string-course and gargoyles beneath embattled ashlar parapet with 8 crocketed pinnacles.

GRADE: I

HERITAGE LIST NUMBER: 1151457

266 Campsall St Mary Magdalene

LOCATION: The Avenue, Campsall

POSTCODE: DN6 9LG

OS GRID REF: SE544141

DATE: Norman

DESCRIPTION: West buttresses (that on right with moulded band and large offsets) flank restored C12 doorway with shafts to 4-order round arch, scalloped capitals and zigzag ornament; above, a blind arcade of 5 arches divided by shafts with block capitals, the central light taller and with zigzag; string-course beneath 1-light window with shafts, clock above; offset beneath belfry stage with small blind lights flanked by louvred 2-light openings with shafts and with round arch over; string-course beneath Perpendicular embattled parapet with truncated corner pinnacles.

GRADE: I

HERITAGE LIST NUMBER: 1151464

267 Cantley St Wilfrid

LOCATION: Church Lane, Cantley

POSTCODE: DN4 6QR

OS GRID REF: SE619015

DATE: C15

DESCRIPTION: 3-stage ashlar west tower with diagonal offset buttresses up to bell stage to all corners and deep moulded plinth, also string-course between stages. Deeply chamfered 4-centred arched west door with Perpendicular-traceried 3-light window over, in deep 4-centred arched surround with original hood-mould with head stops. Blind second stage. Bell stage with pointed 2-light cusped traceried openings in deep cavetto surrounds plus original hoods. Eaves string-course with central and corner gargoyles, embattled parapets with crocketed corner and central pinnacles.

GRADE: II*

HERITAGE LIST NUMBER: 1151446

268 Conisbrough St Peter

LOCATION: Church Street/High Street, Conisbrough

POSTCODE: DN12 3HL

OS GRID REF: SK512987

DATE: C15

DESCRIPTION: Saxon core refaced C15; much restored. Chamfered plinth, moulded band; offset angle buttresses flank pointed-arch west door with hood-mould; string-course beneath pointed-arch 3-light window having unrestored hood-mould with head-carved stops; statue niche over. North and south clocks and offset beneath belfry stage having pairs of transomed, 2-light openings with cusped blind panels below louvres; shared hood-moulds with head-carved stops. String-course with north and south gargoyles and traceried frieze beneath embattled parapet with 4 crocketed pinnacles.

GRADE: I

HERITAGE LIST NUMBER: 1192787

269 Darfield All Saints

LOCATION: Church Street, Darfield

POSTCODE: S73 9JX

OS GRID REF: SE419044

DATE: C11 and C12 with C15 upper stages

DESCRIPTION: West tower of 4 stages, quoins to lower 2. Restored west doorway with hood-mould; a 3-light Perpendicular window above; several small slit windows to right; a lancet to each return. Ashlar upper stages set back at string-courses. Clock beneath a pair of 2-light belfry openings on west side, 3-light belfry opening to each return. Cavetto-moulded course with waterspouts beneath embattled parapet with corner pinnacles.

GRADE: I

HERITAGE LIST NUMBER: 1315015

270 Fishlake St Cuthbert

LOCATION: Church Lane, Fishlake

POSTCODE: DN7 5JW

OS GRID REF: SE656132

DATE: C13–C15

DESCRIPTION: Tower of 4 stages. Chamfered plinth and 2 moulded bands; offset angle buttresses. C13 west door with pointed arch and hood-mould beneath large, transomed, 5-light window with cusped lights, traceried head and hood-mould. Above south aisle are carvings of a bird and crown. Third stage has west niche with statuette of St Cuthbert beneath a crocketed pinnacle. String-course below paired 2-light belfry openings with louvres, hood-moulds and dividing buttresses. Iron-faced west clock across string-course beneath embattled parapet with 8 crocketed pinnacles; gargoyles to north and south.

GRADE: I

HERITAGE LIST NUMBER: 1314801

271 Harthill All Hallows

LOCATION: Union Street, Harthill

POSTCODE: S26 7YZ

OS GRID REF: SK494810

DATE: C15

DESCRIPTION: Perpendicular-style tower with chamfered plinth and moulded band; stair-turret door on south side. Offset angle buttresses flank C19 3-light west window with head-stopped hood-mould. String-courses below and above west clock; pointed 2-light belfry openings with louvres and hood-moulds. Gargoyles to north and south beneath embattled parapet with crocketed pinnacles.

GRADE: I

HERITAGE LIST NUMBER: 1132709

272 Hickleton
St Wilfrid

LOCATION: Barnsley Road, Hickleton

POSTCODE: DN5 7BA

OS GRID REF: SE483053

DATE: Late C15–early C16

DESCRIPTION: West tower with chamfered plinth and moulded band, diagonal offset buttresses. 3-light west window displaced to left by internal stair-turret. Clock beneath 2-light belfry openings with renewed, uncusped tracery. String-course with gargoyles, saddleback roof behind parapet.

GRADE: I

HERITAGE LIST NUMBER: 1314784

273 High Melton
St James

LOCATION: Church Lane, High Melton

POSTCODE: DN5 7SY

OS GRID REF: SE509019

DATE: Late C14–C15

DESCRIPTION: Perpendicular tower. Offset diagonal buttresses flank 2-light west window with ogee-headed lights and hood-mould; slit windows to stair-turret in north-west buttress. Small ogee-headed lights and string-course beneath second-stage clock. String-course beneath later upper stage with transomed, 2-light belfry openings. String-course with corner gargoyles beneath embattled parapet with crocketed pinnacles.

GRADE: II*

HERITAGE LIST NUMBER: 1192208

274 Hooton Pagnell All Saints

LOCATION: Adjacent to Hooton Pagnell Hall

POSTCODE: DN5 7BS

OS GRID REF: SE485080

DATE: C14

DESCRIPTION: West tower: rubble lower stage with diagonal, ashlar south-west and angle north-west buttresses; C19 west window of 3 lights, reticulated tracery. Moulded offset beneath 2 C14 ashlar stages, south and north clock faces, 2-light belfry openings with louvres and cusped Y-tracery. String-course with gargoyles; embattled, pinnacled parapet.

GRADE: I

HERITAGE LIST NUMBER: 1314808

275 Kirk Bramwith St Mary

LOCATION: Top Lane, Kirk Bramwith

POSTCODE: DN7 5SW

OS GRID REF: SE620118

DATE: C14–C15

DESCRIPTION: Tower with offset and moulded bands and diagonal west buttresses (that to north-west with carved cross). Plain 2-light west window with slit to stair-turret on left. Cruciform window and string-course beneath C15 upper stage with chamfered 2-light windows; string-course beneath crenellated parapet with crocketed corner.

GRADE: II*

HERITAGE LIST NUMBER: 1286522

276 Kirk Smeaton St Peter

LOCATION: Main Street, Kirk Smeaton

POSTCODE: WF8 3JY

OS GRID REF: SE520166

DATE: C14

DESCRIPTION: The 2-stage west tower has diagonal buttresses with offsets to first stage; chamfered plinth; to west side a 2-light window with recut Y-tracery within chamfered surround, further 2-light window with reticulated-type tracery to head within double-chamfered surround. Chamfered band, pointed 2-light bell-opening with hood-mould to each side. Moulded string-course with gargoyles. Battlements with crocketed finials.

GRADE: II*

HERITAGE LIST NUMBER: 1316345

277 Loversall St Katherine

LOCATION: Bubup Hill, turn by church noticeboard

POSTCODE: DN11 9DD

OS GRID REF: SK576988

DATE: C14–C15

DESCRIPTION: Tower of 4 stages; lower part c.1300, altered; top stage Perpendicular. Chamfered plinth, moulded band, diagonal west buttresses and string-courses between stages. Pointed west door, offset to left, has quadrant-moulded surround as does 3-light window over; Perpendicular 2-light window to third stage. Offset beneath belfry stage with louvred, 2-light pointed windows and string-course with north and south gargoyles beneath embattled parapet with 8 crocketed pinnacles.

GRADE: II*

HERITAGE LIST NUMBER: 1193085

278 Norton St James

LOCATION: Norton Church Road, Norton

POSTCODE: S8 8JQ

OS GRID REF: SK359821

DATE: C13

DESCRIPTION: The west tower of 3 stages, has a moulded plinth, string-courses, angle buttresses topped with pinnacles, crenellated parapet and gargoyles. First stage has to west a 3-light pointed-arch panel-tracery window with transom and hood-mould. Second stage has to south a single lancet with a clock above. Bell stage has a C13-style 2-light pointed-arch bell-opening on each side, with hood-mould.

GRADE: II*

HERITAGE LIST NUMBER: 1271180

279 Old Edlington St Peter

LOCATION: B6376, approx 0.5 mile south of New Edlington

POSTCODE: DN12 1PX

OS GRID REF: SK532972

DATE: C12–C15

DESCRIPTION: Tower has altered C12 base and Perpendicular belfry stage. Offset diagonal west buttresses have chamfered plinths, moulded bands and carved heads near their tops. Perpendicular 2-light west window with head-carved hood-mould stops. Scalloped band and string-course beneath belfry stage having unlouvred, 2-light pointed openings with hood-moulds; string-course with north and south gargoyles beneath embattled parapet with 4 crocketed pinnacles.

GRADE: I

HERITAGE LIST NUMBER: 1286332

280 Owston All Saints

LOCATION: Stockbridge Lane, Owston

POSTCODE: DN6 9JF

OS GRID REF: SE551111

DATE: Early C13

DESCRIPTION: Tower of 3 stages. Chamfered plinth, offsets between stages. Round-headed loops to lower stage on west and south sides and to second stage on west; louvred belfry openings each of 2 pointed lights divided by shaft with carved capital set in a pointed-arch recess; string-course with corner gargoyles beneath C15 embattled parapet with 8 crocketed pinnacles.

GRADE: I

HERITAGE LIST NUMBER: 1192336

281 Rossington St Michael

LOCATION: Sheep Bridge Lane, Rossington

POSTCODE: DN11 0EZ

OS GRID REF: SK625984

DATE: Mid-C14

DESCRIPTION: Perpendicular tower of 4 stages; magnesian limestone. Chamfered plinth and moulded band; diagonal west buttresses with offsets; string-courses between stages. Small west doorway in quadrant-moulded surround with hood-mould beneath 3-light window with plain tracery in similar surround; clock faces to short third stage. Pointed 2-light belfry openings with louvres, panel tracery and hood-moulds; north and south gargoyles to embattled parapet with 8 crocketed pinnacles.

GRADE: II*

HERITAGE LIST NUMBER: 1151519

282 Snaith St Lawrence

LOCATION: High Street, Snaith

POSTCODE: DN14 9HE

OS GRID REF: SE641222

DATE: C13

DESCRIPTION: Unbuttressed 4-stage tower: stair-lighting slits to south-west angle, string-courses between stages. First stage: pointed west door with 3 chamfered and one inner keeled order on restored shafts with moulded capitals. Second stage: west lancet with chamfered and keeled arch and hood-mould on twin shafts, north and south lancets. Third stage: single lancets to each side, clock face to south. Belfry: 3 lancets to each side with double-chamfered reveals and hood-moulds. Corbel-table with weathered head corbels; embattled parapet with 8 pinnacles.

GRADE: I

HERITAGE LIST NUMBER: 1161899

283 Sprotbrough St Mary

LOCATION: Main Street, Sprotbrough

POSTCODE: DN5 7LP

OS GRID REF: SE540020

DATE: C14

DESCRIPTION: Lower part has chamfered plinth and diagonal offset buttresses flanking ogee-headed west door with hood-mould; quadrant-moulded 2-light window above with trefoil-headed loop over. Offsets below and above lozenge-shaped clock. Perpendicular upper stage has paired, 2-light belfry openings with cusped, blind panels beneath louvred upper lights, shared hood-moulds. Blind tracery to parapet with gargoyles and pinnacles.

GRADE: I

HERITAGE LIST NUMBER: 1192663

284 Thorne
St Nicholas

LOCATION: Stonegate, Thorne

POSTCODE: DN8 5NP

OS GRID REF: SE690132

DATE: C13

DESCRIPTION: Tall lower stage with later buttresses flanking a 2-light, Y-tracery west window with hood-mould; short second stage with string-course below and above Y-tracery window of 2 lights; Perpendicular belfry stage has paired 2-light openings with louvres and shared hood-moulds; embattled parapet with crocketed pinnacles, west clock, flagpole and weathervane.

GRADE: I

HERITAGE LIST NUMBER: 1193076

285 Thorpe Salvin
St Peter

LOCATION: Harthill Road, centre of Thorpe Salvin village

POSTCODE: S80 3JR

OS GRID REF: SK521811

DATE: C12, heightened in C15

DESCRIPTION: Chamfered plinth and moulded band; buttresses to east, that on north side extended as a stack. 3-light west window, offset to north, has cusped pointed lights in hollow-chamfered surround with hood-mould; slit windows to internal stair on right. String-course and offset beneath pointed 2-light belfry openings with louvres and hood-moulds. String-course with central gargoyles beneath embattled parapet with crocketed corner pinnacles.

GRADE: I

HERITAGE LIST NUMBER: 1314667

286 Tickhill
St Mary

LOCATION: St Mary's Road, Tickhill

POSTCODE: DN11 9LY

OS GRID REF: SK591931

DATE: Early C13–C15

DESCRIPTION: West tower which can be dated by historic evidence 1373–99. Tower still incomplete in 1429, possibly providing a date for the crenellated parapet. Lower stage of tower, which can be dated to early C13, includes simple clasping buttresses, and elaborate Early English west doorway, all decorated with nail head.

GRADE: I

HERITAGE LIST NUMBER: 1151698

287 Todwick
St Peter & St Paul

LOCATION: Lindleys Croft, Todwick

POSTCODE: S26 1HN

OS GRID REF: SK498841

DATE: C15

DESCRIPTION: 3-stage tower in Perpendicular style. Chamfered plinth and moulded band. Diagonal buttresses flank 3-light west window with cusped, pointed lights in cavetto-moulded surround with pointed arch and hood-mould. String-courses between stages, slit windows to south-west internal stair. 2-light louvred belfry openings with pointed arches and hood-moulds. North and south waterspouts beneath embattled parapet with 8 crocketed pinnacles.

GRADE: II*

HERITAGE LIST NUMBER: 1151903

288 Treeton St Helen

LOCATION: Front Street/Church Lane, Treeton

POSTCODE: S60 5PZ

OS GRID REF: SK432877

DATE: C12–C15

DESCRIPTION: Rubble walling to C12 lower part of tower; lancet west window, string-course with 3 slit windows set inline above; clock on east side. Weathered string-course beneath Perpendicular upper stage with transomed 2-light belfry openings, string-course with waterspouts beneath embattled parapet with crocketed pinnacles.

GRADE: I

HERITAGE LIST NUMBER: 1132728

289 Wadworth St John Baptist

LOCATION: Church Road, Wadworth

POSTCODE: DN11 9BW

OS GRID REF: SK569970

DATE: C14–C15

DESCRIPTION: Perpendicular tower; chamfered plinth and moulded band. Diagonal west buttresses flank pointed 3-light west window with hood-mould; north and south clocks beneath belfry stage having pairs of transomed 2-light windows with blind, cusped panels below louvres and shared hood-moulds. String-courses below and above traceried frieze; gargoyles beneath embattled parapet with 8 crocketed pinnacles.

GRADE: I

HERITAGE LIST NUMBER: 1151504

290 Wales
St John Baptist

LOCATION: Church Street, Wales

POSTCODE: S26 5LQ

OS GRID REF: SK478827

DATE: C15

DESCRIPTION: Tower of 3 stages; Perpendicular style. Chamfered plinth and moulded band. Diagonal buttresses flank pointed-arch 3-light west window with hood-mould; string-courses below and above clock; pointed 2-light louvred belfry openings; north gargoyle beneath altered parapet, pitched roof.

GRADE: II*

HERITAGE LIST NUMBER: 1286360

291 Wentworth
Holy Trinity Old Church

LOCATION: Church Drive, Main Street, Wentworth

POSTCODE: S62 7TX

OS GRID REF: SK384984

DATE: C14–C15

DESCRIPTION: Chamfered plinth. Small west window offset to left: 3 cusped lights, 4-centred arch, hood-mould. Quatrefoil in chamfered, square opening above and to right. String-course beneath blocked 2-light belfry openings with depressed 4-centred arches, hood-moulds. String-course with corner gargoyles.

GRADE: II*

HERITAGE LIST NUMBER: 1314580

292 Whitgift St Mary Magdalene

LOCATION: Main Street, Whitgift

POSTCODE: DN14 8HQ

OS GRID REF: SE808227

DATE: C14 lower section with C15–C16 upper stages

DESCRIPTION: 3-stage tower (the bottom now below ground level): moulded plinth, full-height angle buttresses with offsets, moulded string-courses between stages. Tall first stage has blocked C14 pointed west door with 2 wave-moulded orders, pointed 4-light transomed west window with round-arched lights, plain Perpendicular tracery and hood-mould. Second stage has slit light to south, clock face to north dated 1919. Top stage has 4-centred arched 2-light belfry openings with cinquefoiled lights and Perpendicular tracery. String-course, coped embattled parapet with crocketed angle pinnacles and plain replacement pinnacles to centres of each side.

GRADE: I

HERITAGE LIST NUMBER: 1083151

293 Wickersley St Alban

LOCATION: Church Lane, Wickersley

POSTCODE: S66 1ES

OS GRID REF: SK478917

DATE: C15

DESCRIPTION: Tower of 4 stages, lower 3 stages C15 in Perpendicular style, upper stage added C19 in same style. Chamfered plinth and moulded band; pointed-arch door in south side. Offset diagonal buttresses flank 3-light west window with hood-mould. String-courses to each stage, third stage has blocked belfry window of 2 cusped lights with hood-mould. Tall C19 upper stage with blocked 2-light west window; north and south windows of both C15 and C19 belfry stages are louvred. String-course with corner gargoyles and waterspouts beneath embattled parapet with 8 crocketed pinnacles.

GRADE: II*

HERITAGE LIST NUMBER: 1193491

Diocese of

Southwell and Nottingham

294 Arnold St Mary

LOCATION: Church Lane, Arnold

POSTCODE: NG5 8HG

OS GRID REF: SK586461

DATE: C14 and mid-C15

DESCRIPTION: West tower, C14 and mid-C15, 2 stages, has moulded plinth, string-course, eaves band and plain parapet with 4 pinnacles. 4 pairs of corner buttresses, 6 set-offs. First stage has to west moulded doorway with clustered shafts and hood-mould. Above, restored triple lancet with geometrical tracery and hood-mould. Second stage has to south carved panel, C13, with figures. West side has square opening. North, south and west sides each have a clock. Each side has, above, a double bell-opening with segmental head.

GRADE: II*

HERITAGE LIST NUMBER: 1235987

295 Askham St Nicholas

LOCATION: Town Street, Askham

POSTCODE: NG22 0RU

OS GRID REF: SK740750

DATE: C15

DESCRIPTION: The C15 diagonally buttressed tower of 2 stages with a band at the juncture is set on a plinth with a moulded band over. Embattled with 8 crocketed pinnacles it has 2 gargoyles to both the north and south sides. The west side has a single-arched window with 3 cusped lights. There are 4 arched bell-chamber openings each with 2 arched lights and panel tracery. There are single rectangular stair lights to the west and north and 2 to the south. The east side has a single clock face.

GRADE: II*

HERITAGE LIST NUMBER: 1045707

296 Averham St Michael & All Angels

LOCATION: Church Lane, Averham

POSTCODE: NG23 5RB

OS GRID REF: SK768543

DATE: C11, C13 and C15

DESCRIPTION: The west tower of 2 stages, C11, C13 and C15, has a chamfered plinth, coved string-course, coved eaves band with 8 Romanesque-style gargoyles, crenellated parapet with shields in panels and 8 crocketed pinnacles. 4 diagonal buttresses, 5 and 3 set-offs, those to east engaged with nave. That to south-west has rebus with moulded hood. First stage has to south, a clock. To west, C14 double lancet with cusped heads and Decorated tracery, chamfered mullion, chamfered and rebated reveal, and hood-mould. Above, a clock. Second stage has 4 C14 double-lancet openings with cusped heads, Decorated tracery, coved mullions, chamfered and rebated reveals with hood-moulds.

GRADE: I

HERITAGE LIST NUMBER: 1046008

297 Babworth All Saints

LOCATION: Babworth Road, Babworth

POSTCODE: DN22 8EP

OS GRID REF: SK686808

DATE: C15

DESCRIPTION: The diagonally buttressed tower of 2 stages with string-course at the juncture has a restored arched 3-light west window with panel tracery and cusping. There is a hood-mould over and string-course under. The bell-chamber has 4 arched 2-light openings with cusping and hood-moulds. There are small single fixed lights to the west and south sides with a clock face under the south belfry opening. On the north and south sides are 2 gargoyles with single crocketed pinnacles at the angles of the parapet.

GRADE: I

HERITAGE LIST NUMBER: 1370370

298 Barnby in the Willows All Saints

LOCATION: Front Street, Barnby in the Willows

POSTCODE: NG24 2SA

OS GRID REF: SK860521

DATE: C15

DESCRIPTION: The diagonally buttressed C15 single-stage tower has a string-course over the plinth and a single-arched 2-light window with tracery to the west with four 2-light arched openings to the bell-chamber. There is a single small rectangular window to the west side with 2 to the south. The tower is embattled with 4 crocketed pinnacles and guttering running under.

GRADE: I

HERITAGE LIST NUMBER: 1302715

299 Beckingham All Saints

LOCATION: Church Street, Beckingham

POSTCODE: DN10 4PH

OS GRID REF: SK779902

DATE: C13–C15

DESCRIPTION: West tower of 4 stages; plinth, 4 string-courses, crenellated parapet, 8 crocketed pinnacles, has 2 diagonal buttresses, 4 set-offs plus plinth. West door has moulded pointed-arch head with hood-mould, and close-boarded door. Fourth stage has double-lancet bell-openings with panel tracery, chamfered reveals, hood-moulds and clock faces to south and east.

GRADE: II*

HERITAGE LIST NUMBER: 1045129

300 Bilborough St Martin of Tours

LOCATION: St Martin's Road, Bilborough

POSTCODE: NG8 3BH

OS GRID REF: SK524418

DATE: c.1450

DESCRIPTION: Square west tower, 2 stages, has angle buttresses to the lower stage, and crenellated parapet with string-course. To west, a pointed-arch 2-light window with transom and tracery. Above, single loops on each side. Bell stage has a flat-headed 2-light opening on each side.

GRADE: II*

HERITAGE LIST NUMBER: 1255110

301 Bilsthorpe St Margaret

LOCATION: Church Hill, Bilsthorpe

POSTCODE: NG22 8RU

OS GRID REF: SK655602

DATE: C14 and 1663

DESCRIPTION: West tower of 2 stages, has chamfered plinth, string-course, crenellated parapet, 2 obelisk pinnacles to west. 2 diagonal buttresses, 3 set-offs. First stage has, to west, late C14 cusped triple lancet in coved 4-centred arched reveal with hood-mould and stops. Above, remains of clock. Second stage has 4 double round-headed openings, 1663, and to south, tablet inscribed "... Chwardens 1663".

GRADE: I

HERITAGE LIST NUMBER: 1045616

302 Blyth
St Mary & St Martin

LOCATION: Bawtry Road, opposite Red Hart pub, Blyth

POSTCODE: S81 8HL

OS GRID REF: SK624873

DATE: C15

DESCRIPTION: The angle-buttressed C15 tower of 2 stages has a chamfered base with 2 moulded string-courses over. There is a central arched west doorway with moulded jambs and arch, with inner order of fleuron decoration. Over is a hood-mould decorated with worn crockets and rising to a crocketed pinnacle with decorated finial. Above is a slightly projecting canopy. The outer order of the jambs support single slender flat piers with set-offs. These are decorated with blind tracery and topped with a crocketed pinnacle and finial which terminate at the canopy. Above is an arched 3-light window divided by a single transom and 3 mullions, with reticulated tracery below the transom and reticulated and panel tracery above. Over is a hood-mould and label stops. On either side are single niches, having ogee arches topped with a crocketed pinnacle. Rising from the springing of these arches are single small slim piers with worn decorated capitals. Above the window is a similar niche and above a single square light. There are 6 small stair lights. The 4 large arched bell-chamber openings of 4 arched lights, surmounted by a further 4 arched lights and tracery, have a hood-mould with finial and label stops, the latter rising to single gargoyles. The top of the tower is embattled with centre and corner crocketed pinnacles. The merlons are conjoined by single openwork crocketed and cusped arches.

GRADE: I

HERITAGE LIST NUMBER: 1239182

303 Bole
St Martin

LOCATION: East Street, Bole

POSTCODE: DN22 9ER

OS GRID REF: SK783871

DATE: C15

DESCRIPTION: West tower, 1504 ashlar, 3 stages on plinth, 2 string-courses, battlemented parapet, 8 crocketed pinnacles, 2 diagonal buttresses to west, 3 set-offs; plain buttresses to north-east and south-east, 4 set-offs that to north engaged with aisle. West window C15 cove-moulded surround containing 3 round-headed lights with chamfered mullions and 4 similar lights above. Third stage has 4 double-lancet bell-openings with cusped heads; C19 iron clock dial to west.

GRADE: I

HERITAGE LIST NUMBER: 1045083

304 Car Colston
St Mary

LOCATION: Screveton Road, Car Colston

POSTCODE: NG13 8JF

OS GRID REF: SK720430

DATE: C13 and C14

DESCRIPTION: West tower, 2 stages, has deep moulded plinth, string-course and coved eaves with 2 gargoyles each side. Crenellated parapet with 4 crenellated square pinnacles and 4 intermediate crocketed pinnacles. Squat set-back octagonal spire with crocketed finial and weathervane. To west, 2 diagonal buttresses, 2 set-offs. To east, pair of plain buttresses. First stage has, to south, small C13 trefoiled lancet and clock. To west, a chamfered lancet and above, trefoiled lancet and clock. Second stage has 4 transomed C15 double lancets with panel tracery and hood-moulds.

GRADE: I

HERITAGE LIST NUMBER: 1045667

305 Carlton in Lindrick St John the Evangelist

LOCATION: Church Lane, Carlton in Lindrick

POSTCODE: S81 9EG

OS GRID REF: SK588839

DATE: C11 and C15

DESCRIPTION: The C11 tower was diagonally buttressed between 1417 and 1443, given a south-west staircase and topped with an embattled belfry with 8 crocketed pinnacles and single central gargoyle to each side. The C11 part is in 4 stages marked respectively from the base by coursed rubble, roughly dressed rectangular blocks, herringbone and coursed rubble interspersed with herringbone. There are 2 string-courses, one square in section, the second at the junction of the C11 and C15 stages is C15. The west mid-C12 doorway was moved here from the south in 1831. The inner jambs and arch around the double wooden door are decorated with chevrons. Either side are 3 engaged columns with volute capitals. The inner supports an arch decorated with chevrons on front and soffit separated by a narrow band of pointed roll-moulding. The outer 2 support moulded arches. There is a hood-mould with head label stops. Above is an opening with 2 cusped, arched lights under a flat head with hood-mould over. To its right is a small rectangular stair light. Above on west and east sides are single coupled C11 bell-openings. The lintels are carved to form 2 arches which are separated by a plain shaft. In the C15 belfry are four 3-light arched openings with cusping and hood-mould. To the right of the west opening is a small rectangular stair light. Under the north and south openings are single C19 clock faces. Abutting the south-west buttress is the base of the C14 church cross. Abutting the south and north aisle walls are single-coursed rubble C12 nave quoins. The C19 buttressed north wall with embattled parapet is set on a plinth.

GRADE: I

HERITAGE LIST NUMBER: 1045742

306 Caunton St Andrew

LOCATION: Main Street, Caunton

POSTCODE: NG23 6AA

OS GRID REF: SK745600

DATE: C13–C15

DESCRIPTION: The west tower has a moulded plinth and 4 diagonal buttresses with 3 offsets to the west and 2 to the east. It is in 2 stages, marked by a string-course. The first stage has, to the south, 3 stair lights and a small slit window; and to the west, a C13 roll-moulded doorway with hood-mould, and above a C15 triple lancet with cusped heads in a coved 4-centred arched surround with splayed mullions. The stained glass was inserted in 1905 and depicts 3 saints, each with an angel above. Each face of the tower has a C19 clock. The second stage has 4 C14 cusped trefoil-headed, double-lancet bell-openings with splayed mullions, in chamfered and splayed openings with hood-moulds. Above is a panelled eaves band with 4 gargoyles and a moulded crenellated parapet.

GRADE: I

HERITAGE LIST NUMBER: 1045974

307 Church Warsop
St Peter & St Paul

LOCATION: Church Street, Church Warsop

POSTCODE: NG20 0SN

OS GRID REF: SK568688

DATE: C11 and C14

DESCRIPTION: West tower, 3 stages, has clasping and corner buttresses of different sizes, some gabled. String-course, moulded crenellated parapet, 2 gargoyles, 4 crocketed pinnacles. First stage has to south additional buttress, 2 set-offs. To west, restored C11 round-headed door with imposts. Above, restored round-headed window. Second stage has to south a round-headed window and 2 stair lights. Above, clock, 1844, to south and east. Third stage has 4 C14 double-lancet bell-openings with cusped heads.

GRADE: I

HERITAGE LIST NUMBER: 1240271

308 Clarborough
St John the Baptist

LOCATION: Church Lane, Clarborough

POSTCODE: DN22 9NQ

OS GRID REF: SK734833

DATE: C13

DESCRIPTION: 2-stage west tower has a pair of diagonal buttresses, eaves band, 4 gargoyles, crenellated parapet with 4 crocketed pinnacles. West front has C19 double lancet with hood-mould. Second stage has 4 C13 cusped double-lancet bell-openings with chamfered reveals and hood-moulds.

GRADE: I

HERITAGE LIST NUMBER: 1370125

309 Clayworth St Peter

LOCATION: Town Street, Clayworth

POSTCODE: DN22 9AB

OS GRID REF: SK726884

DATE: C14 and C15

DESCRIPTION: West tower, 2 stages, has plinth, string-course, eaves band with masks, 8 gargoyles, crenellated parapet with 8 crocketed pinnacles. West side has restored late C14 triple lancet with panel tracery and hood-mould with mask stops. Above, single small lancet. Above, to south, clock. Second stage has 4 C15 double-lancet bell-openings with cusped quatrefoil heads, moulded reveals and coved hood-moulds.

GRADE: I

HERITAGE LIST NUMBER: 1212157

310 Collingham All Saints

LOCATION: Low Street/Church Lane, Collingham

POSTCODE: NG23 7NF

OS GRID REF: SK830620

DATE: C13

DESCRIPTION: West tower, 2 stages, battlemented parapet with 4 crocketed pinnacles, 2 diagonal buttresses to west with 3 set-offs plus plinth. West window has C13 moulded opening with label mould containing triple lancet with restored panel tracery; above, clock, and 2 round-headed openings on south. Second stage has triple-lancet bell-openings with 4-centred arched heads.

GRADE: I

HERITAGE LIST NUMBER: 1156985

311 Collingham St John the Baptist

LOCATION: Church Street, Collingham

POSTCODE: NG23 7LN

OS GRID REF: SK824613

DATE: C13 and C14

DESCRIPTION: West tower, 3 stages, string-courses and corbel-table with crenellated parapet and 4 crocketed pinnacles. 3 clasping buttresses of 6 set-offs, changing to diagonal buttresses above first stage. North-west corner only has plain buttress of 3 set-offs up to first stage. First stage has, to west, C14 chamfered and rebated 4-centred arched window with label mould and mask stops, containing triple ogee-headed lancet with panel tracery; to south, single lancet in chamfered reveal; to north, 1890 polygonal stair-turret with pitched slab roof and 3 rectangular lights with chamfered reveals. Second stage has, to west, single lancet in chamfered reveal in blocked C13 double-lancet opening; to south, C13 double lancet; to north, double recess with one lancet and one round-headed opening; to east, clock. Third stage has 4 triple-lancet bell-openings with moulded mullions and transoms and Decorated tracery.

GRADE: I

HERITAGE LIST NUMBER: 1046050

312 Cromwell St Giles

LOCATION: Main Street, Cromwell

POSTCODE: NG23 6JD

OS GRID REF: SK799615

DATE: c.1427

DESCRIPTION: West tower, c.1427, 4 stages. Chamfered plinth, 4 string-courses, eaves band with 6 gargoyles, crenellated parapet. At north-east corner, a full-height stair-turret. First stage has, to west, splayed doorway with hood-mould and mask stops; to north, a stair light. Second stage has, to west, a C15 triple lancet with cusped ogee heads, splayed mullion, chamfered and rebated reveal and hood-mould; to north, a stair light. Third stage has, to west, a cusped-head lancet in splayed and rebated square-headed reveal with hood-mould; to north, stair lights. Fourth stage has 4 C15 cusped-head double lancets with Perpendicular tracery, coved mullions, coved and splayed reveal and hood-mould.

GRADE: I

HERITAGE LIST NUMBER: 1369982

313 Cropwell Bishop St Giles

LOCATION: Fern Road, Cropwell Bishop

POSTCODE: NG12 3BU

OS GRID REF: SK684355

DATE: c.1450

DESCRIPTION: Perpendicular tower of c.1450 with clasping and angle buttresses, in 3 reducing stages. 3-light west window, small 2-light bell-chamber openings to top stage, double decorative frieze, crenellated parapet with corner pinnacles.

GRADE: I

HERITAGE LIST NUMBER: 1045650

314 Dunham on Trent St Oswald

LOCATION: Main Street, Dunham on Trent

POSTCODE: NG22 0TY

OS GRID REF: SK815745

DATE: C15

DESCRIPTION: The C15 tower with embattled parapet and 4 crocketed pinnacles has 2 gargoyles on the north side and the remains of single gargoyles to both the west and south sides. The diagonal buttresses are gabled part way up where they are decorated with crockets. There are single string-courses extending over the plinth and at the juncture of each of the 4 stages. The west is of 3 stages. West doorway with pointed-arch surround, hood-mould, label stops and wooden door. Above is a single ogee-arched and vaulted niche with 2 bosses, the arch decorated with crockets; over is a hood-mould with grotesque head label stops and ornate finial. Above is a single large moulded arched window with 4 lights, panel tracery, cusping, hood-mould, figurative label stops and finial. There are 4 very large arched and traceried belfry openings each with 4 arched and cusped lights surmounted by a further 4 similar lights. There are 3, 4 and single small rectangular stair lights to the west, south and north sides respectively. The east side has evidence of a former nave roof.

GRADE: I

HERITAGE LIST NUMBER: 1370101

315 Eakring St Andrew

LOCATION: Kirklington Road, Eakring

POSTCODE: NG22 0DA

OS GRID REF: SK676621

DATE: C13 and C15

DESCRIPTION: West tower, C13 and C15, 3 stages, has 2 string-courses, moulded eaves with remains of gargoyles, crenellated parapet and 4 crocketed pinnacles. First stage has, to south-east, 2 plain buttresses, single set-off, engaged with nave; to west, chamfered C13 door with hood-mould and stops; above, remains of a figure carving and chamfered single lancet with trefoil head and hood-mould. North side has, to east, a plain buttress, single set-off. Second stage has 4 restored double lancets with cusped ogee heads. Third stage has 4 double-lancet louvred openings, that to west with cusped round heads and quatrefoil in circle. Clock faces to north, south and east.

GRADE: II*

HERITAGE LIST NUMBER: 1370132

316 East Drayton St Peter & St Paul

LOCATION: North Green, East Drayton

POSTCODE: DN22 0LH

OS GRID REF: SK775753

DATE: C13 and C15

DESCRIPTION: Diagonally buttressed tower set on a plinth with 2 heavily moulded bands over, of 4 stages with a string-course at each juncture. The parapet has 7 remaining crocketed pinnacles of 8 with 8 gargoyles under. West doorway with moulded pointed arch, hood-mould and remnants of label stops. Above is a single-arched 3-light window with hood-mould and worn label stops. In the bell-chamber are 4 arched 2-light openings with panel tracery. There are single clock faces to the south and east sides and 3 small rectangular stair lights to the south.

GRADE: I

HERITAGE LIST NUMBER: 1212946

317 East Markham St John the Baptist

LOCATION: Church Street, East Markham

POSTCODE: NG22 0SA

OS GRID REF: SK743726

DATE: C15

DESCRIPTION: C15 angle-buttressed tower of 3 stages with bands, 8 crocketed pinnacles and 8 gargoyles. West restored moulded arched doorway with wooden door and continuous hood-mould. Above is a single-arched window with 3 cusped lights surmounted by a further 3 arched and cusped lights with cusped panel tracery, hood-mould and label stops. The south side has at the second stage a niche. Within a cinquefoiled canopy with 3 finials is a standing, worn figure. The 4 arched bell-chamber openings each have 2 arched and cusped lights surmounted by a further 2 similar lights with tracery, hood-mould and label stops. Under the north opening is a single clock face. There are 5 rectangular stair lights to the south side and single similar lights to the west and north sides.

GRADE: I

HERITAGE LIST NUMBER: 1223679

318 Egmanton Our Lady of Egmanton

LOCATION: Tuxford Road, Egmanton

POSTCODE: NG22 0EZ

OS GRID REF: SK736689

DATE: C15

DESCRIPTION: West tower, C15, 3 stages, has moulded and chamfered plinth, 2 string-courses, eaves band with 2 gargoyles to north and south, crenellated parapet with 4 crocketed pinnacles. Partial buttress to south-east, engaged with nave. Pair of diagonal buttresses to west, 3 set-offs. First stage has, to south, a stair light. To west, moulded doorway, early C15, with C15 door. Above, round-headed triple lancet with panel tracery, partly restored, with hood-mould. To its right, a stair light. Second stage has a stair light to east and west. Third stage has 4 C15 double-lancet bell-openings with panel tracery and hood-moulds. Parapet has, on north side, inscription and date 1916.

GRADE: I

HERITAGE LIST NUMBER: 1156553

319 Elston
All Saints

LOCATION: Top Street, Elston

POSTCODE: NG23 5NP

OS GRID REF: SK758480

DATE: C13

DESCRIPTION: Slim C13 tower of 3 stages with band at lower stage, 4 crocketed pinnacles and 8 gargoyles. West side has a single window with 2 arched lights under a flat arch. Above is a single window with 2 arched and cusped lights under a flat arch. The north wall has a doorway with wooden plank door and drip-mould. The 4 arched bell-chamber openings each have 2 arched lights, above are 4 arched windows each with 2 arched lights. There are single rectangular stair lights to the west, north and south. The south side also has a single clock face over the bell-chamber opening.

GRADE: II*

HERITAGE LIST NUMBER: 1370154

320 Everton
Holy Trinity

LOCATION: Church Street, Everton

POSTCODE: DN10 5BB

OS GRID REF: SK691913

DATE: C11–C15

DESCRIPTION: West tower of 3 stages, first 2 rubble with prominent ashlar quoins recut 1983: upper stone ashlar; 2 moulded string-courses, eaves band, moulded crenellated parapet, 2 gargoyles to west. First stage has, on south side, single round-headed unglazed opening; above, clock; to the west, round-headed chamfered window with hood-mould. Third stage has 4 C15 cove-moulded and rebated openings with hood-moulds, each with cusped round-headed double louvred lancet with Decorated tracery.

GRADE: II*

HERITAGE LIST NUMBER: 1156568

321 Farndon
St Peter

LOCATION: West End, Farndon

POSTCODE: NG24 3SG

OS GRID REF: SK768518

DATE: C16

DESCRIPTION: Diagonally buttressed tower, set on a chamfered plinth of 3 stages with bands. Embattled with 8 crocketed pinnacles. The west side has a single window with 3 arched lights under a flat arch with drip-mould. Over this is an ashlar panel inscribed "This was done Ano 1598 …." The south side has a single window with 2 arched lights under a flat arch with flush ashlar quoin surround. Above is a single rectangular stair light. In the bell-chamber are 4 openings, each with 2 arched lights under a flat arch and drip-mould.

GRADE: I

HERITAGE LIST NUMBER: 1178470

322 Finningley
Holy Trinity & St Oswald

LOCATION: Rectory Lane, Finningley

POSTCODE: DN9 3DA

OS GRID REF: SK669991

DATE: Late C11

DESCRIPTION: Tower of 3 stages. Quoins, chamfered plinth; west side has round-headed and rectangular slit windows. String-course beneath louvred belfry openings of 2 round-headed lights divided by shafts with block capitals set in round-arched recesses. Offset string-course beneath embattled ashlar parapet.

GRADE: I

HERITAGE LIST NUMBER: 1286821

323 Fledborough St Gregory

LOCATION: Church Lane, in water meadows at Fledborough

POSTCODE: NG22 0UU

OS GRID REF: SK812722

DATE: C12–C13

DESCRIPTION: The angle-buttressed tower of 2 stages with a string-course at the juncture and pyramidal roof has a pointed chamfered arched west doorway with wooden door. Above is a single-chamfered and deep inner splayed C12 lancet. In the belfry are 4 pointed-arch C13 openings, each with 2 arched lights divided by a single shaft with a lozenge cut into the window heads of the east and west sides and a circle to the north and south sides. There are 2 round tie-plates to the south and north sides.

GRADE: I

HERITAGE LIST NUMBER: 1045689

324 Gamston St Peter

LOCATION: Church Lane, Gamston

POSTCODE: DN22 0QD

OS GRID REF: SK708760

DATE: C14

DESCRIPTION: The C14 tower of 3 stages with bands is set on a deep chamfered plinth with a heavy moulded band over. The angle buttresses rise to diagonal buttresses. The parapet has 8 pinnacles decorated with blind tracery and crocketed finials. There are 2 gargoyles to each side. The west wall has a moulded pointed-arch doorway with heavy panelled door, hood-mould and label stops. Above is a pointed-arch C19 3-light window with cusped panel tracery, hood-mould and label stops. The bell-chamber has 4 arched openings each with 2 blind arched and cusped panels surmounted by a single transom and 2 similar lights with mouchettes over and hood-mould. There is a single rectangular stair light to the north side and 5 similar lights to the south.

GRADE: I

HERITAGE LIST NUMBER: 1224125

325 Granby All Saints

LOCATION: Sutton Lane, Granby

POSTCODE: NG13 9PU

OS GRID REF: SK751362

DATE: C11–C15

DESCRIPTION: The square, 4-stage tower has broad clasping buttresses to the first stage and a single lancet west window. In the second stage are 2 small round-headed Norman windows. 2-light, Y-tracery bell-chamber openings to the third stage. A deep frieze with an ornamental band, crenellated parapet with gargoyles and corner pinnacles forms the Perpendicular top stage.

GRADE: I

HERITAGE LIST NUMBER: 1265230

326 Greasley St Mary

LOCATION: Church Road, Greasley

POSTCODE: NG16 2AB

OS GRID REF: SK490471

DATE: Mid-C15

DESCRIPTION: West tower, mid-C15, 3 stages, has chamfered and moulded plinth, coved string-courses and eaves, 4 gargoyles, crenellated parapet with 8 crocketed pinnacles. 4 pairs of corner buttresses, 5 set-offs. First stage has, to west, moulded doorway with hood-mould and above it, cusped triple lancet with panel tracery and hood-mould. Second stage has a light to south and west. Third stage has on each side 2 double round-headed lancets with panel tracery and hood-mould.

GRADE: II*

HERITAGE LIST NUMBER: 1263847

327 Gringley on the Hill St Peter & St Paul

LOCATION: High Street, Gringley on the Hill

POSTCODE: DN10 4RF

OS GRID REF: SK736906

DATE: C14

DESCRIPTION: West tower, 2 stages, has 2 string-courses, moulded crenellated parapet, 4 crocketed pinnacles. To the west, 2 diagonal buttresses, 2 set-offs plus plinth. To the south, single-chamfered rectangular light. To the west, chamfered C14 doorway with close-boarded door; above, C14 splayed and chamfered opening with coved hood-mould, containing restored C16 triple lancet with panel tracery. To the right, 2 rectangular and one trefoil lights to stair. To the north, blocked round-headed Norman doorway. Second stage has 4 C14 coved and splayed openings with coved hood-moulds containing cusped-head double-lancet louvred bell-openings with panel tracery.

GRADE: II*

HERITAGE LIST NUMBER: 1370395

328 Harworth All Saints

LOCATION: Church Walk, Harworth

POSTCODE: DN11 8LJ

OS GRID REF: SK614913

DATE: C12

DESCRIPTION: Diagonally buttressed C12 tower of 3 stages, set on a plinth with C14 embattled parapet and 8 crocketed pinnacles. On both the north and the south sides are 2 gargoyles. There is a string-course extending over the west window and a further string-course under the bell-chamber openings. The C19 arched 3-light west window with tracery, cusping and hood-mould has to the right a memorial to the Short family, c.1805 with a small stair light over. The south wall has 3 stair lights. There are 4 arched bell-chamber openings, each with 2 lights, tracery and cusping.

GRADE: II

HERITAGE LIST NUMBER: 1045715

329 Hawton
All Saints

LOCATION: Newark Road, Hawton

POSTCODE: NG24 3RN

OS GRID REF: SK788511

DATE: c.1482

DESCRIPTION: The c.1482 tower of 5 stages is set on a plinth with string-course over; there are 4 further string-courses and 4 clasping buttresses. In the west is a doorway with blind flowing traceried door under a moulded arch with hood-mould and label stops. In the spandrel are decorative carved shields and carved heads. Above is a 4-light arched panel-tracery window. To the fourth stage on all but the east wall is an arched window with single mullion and 2 arched transoms. The south has a single small light to the first, second and third stages, with 2 to the fourth stage. There are similar lights to the third, fourth and fifth stages of the west wall, the second stage having 2 lights. The bell-chamber has 4 pairs of 2-light openings with tracery and cusping and a single ogee arch and finial over each pair. The spandrel has blind panel tracery. The parapet is embattled with corner and centre merlons rising to a crocketed pinnacle. Running under is a frieze of shields set in lozenges, those to the west and south-east are decorated. At each corner is a gargoyle.

GRADE: I

HERITAGE LIST NUMBER: 1046031

330 Headon St Peter

LOCATION: Church Street, Headon

POSTCODE: DN22 0RE

OS GRID REF: SK748771

DATE: C13

DESCRIPTION: The angle-buttressed low C13 tower of 2 stages with bands is set on a shallow plinth on the west side and is topped with C19 battlements. The west wall has an arched, recut, 3-light window with intersecting tracery, hood-mould and head label stops. There is evidence of an earlier arched opening over. Above is a single lancet. The north and south sides each have a large blocked chamfered arch with evidence of an earlier arch over flanked by single corbels. The north side has 2 further corbels. Above, to the north and south, are single lancets and above this on the south side is a single rectangular opening.

GRADE: I

HERITAGE LIST NUMBER: 1224198

331 Hockerton St Nicholas

LOCATION: Church Lane, Hockerton

POSTCODE: NG25 0FW

OS GRID REF: SK716564

DATE: C14

DESCRIPTION: Embattled diagonally buttressed tower of 2 stages with bands, upper stage C14, set on a chamfered plinth with moulded band over. Single worn gargoyle on each side. West side has a moulded arched doorway with hood-mould and decorative label stops. Above is a single restored C14 arched 3-light window with cusped panel tracery, hood-mould and human head label stops. Above is a single small blocked trefoil-arched opening. There are 4 arched C14 bell-chamber openings each with 2 arched and cusped lights. The west side has a single and the south side 2 rectangular lights.

GRADE: II*

HERITAGE LIST NUMBER: 1045486

332 Hucknall St Mary Magdalene

LOCATION: Market Place, centre of Hucknall

POSTCODE: NG15 7AS

OS GRID REF: SK533494

DATE: C12 and C14

DESCRIPTION: West tower, late C12, raised C14, 3 stages, has corner pilasters, coved eaves and crenellated parapet with 4 pinnacles. Remains of 2 gargoyles. North and south sides have each a buttress. First stage has, to south, a lancet. To west, restored and partly blocked C13 triple lancet with hood-mould; to its left, C20 gabled shed. Second stage has on each side a paired lancet, C13, with rounded mullion. Third stage has on each side, clock, 1884, and above it, louvred C14 double lancet with Y-tracery.

GRADE: II*

HERITAGE LIST NUMBER: 1217611

333 Kelham St Wilfrid

LOCATION: Averham Road, in the grounds of Kelham Hall

POSTCODE: NG23 5QX

OS GRID REF: SK774553

DATE: C15

DESCRIPTION: West tower, 4 stages, moulded plinth, 3 string-courses, moulded eaves band, 4 gargoyles, 4 crocketed pinnacles. Pair of clasping buttresses to west, changing to diagonal buttresses in third stage. Pair of similar buttresses to east, engaged with aisles. First stage has, to south, a stair light; to west, sill band and a stair light. Second stage has, to west, C16 triple lancet with cusped round heads, Perpendicular tracery, coved mullions, coved, rebated and roll-moulded reveals and hood-mould with mask stops. Third stage has, to south, single elliptical-headed louvred opening; to west, restored chamfered rectangular opening with lattice infill; to north, clock. Fourth stage has 4 C15 double-lancet bell-openings with coved mullions, chamfered reveal and hood-moulds with mask stops.

GRADE: I

HERITAGE LIST NUMBER: 1302213

334 Kirton Holy Trinity

LOCATION: Main Street, Kirton

POSTCODE: NG22 9LU

OS GRID REF: SK691693

DATE: C15

DESCRIPTION: The 2-stage west tower has a coved string-course and eaves band with gargoyles to the north and south. Crenellated parapet with 4 crocketed pinnacles. 2 shallow diagonal buttresses to west, 3 set-offs. First stage has, to south, a stair light and above, clock; to west, late C15 double lancet with round heads, panel tracery and hood-mould, to right, stair light, and above, clock. Second stage has 4 C15 double-lancet bell-openings with round heads, panel tracery and hood-moulds. Nave, 3 bays, has to north gabled octagonal chimney and, to east, a plain buttress, 2 set-offs. South side has, to east, a restored C13 double lancet with plate tracery and hood-mould; to west, late C15 triple lancet with cusped heads and 4-centred arched reveal.

GRADE: II*

HERITAGE LIST NUMBER: 1156898

335 Kneesall St Bartholomew

LOCATION: Newark Road, Kneesall

POSTCODE: NG22 0AE

OS GRID REF: SK704642

DATE: c.1425

DESCRIPTION: West tower, 3 stages, has 2 string-courses and eaves band with bosses and remains of 6 angels and beasts. 2 spouts to north and south. Traceried panelled parapet with 4 restored panelled crocketed pinnacles. Pair of C19 corner buttresses to west, 4 set-offs. Pair of plain buttresses to east, engaged with aisles, 2 set-offs. First stage has, to west, C15 chamfered doorway with hood-mould, mask stops and peak, C19 door. Above, early C15 double lancet, restored, with cusped ogee heads, cusped quatrefoil and mask stops; restored; above again, clock. Third stage has 4 C15 double lancets with cusped ogee heads, cusped quatrefoils, hood-moulds and mask stops.

GRADE: I

HERITAGE LIST NUMBER: 1156936

336 Kneeton St Helen

LOCATION: Church Lane, conspicuous landmark at Kneeton

POSTCODE: NG13 8JS

OS GRID REF: SK709461

DATE: Early C15

DESCRIPTION: West tower of 3 stages, has 2 string-courses, chamfered eaves band, 4 diagonal gargoyles and crenellated parapet. To west, 2 diagonal buttresses, 4 set-offs; to east, 2 plain buttresses. First stage has, to west, a triple lancet with panel tracery, hood-mould and mask stops. Third stage has 4 early C15 double lancets with panel tracery and hood-moulds.

GRADE: II

HERITAGE LIST NUMBER: 1272706

337 Lambley Holy Trinity

LOCATION: Church Street, Lambley

POSTCODE: NG4 4QB

OS GRID REF: SK631454

DATE: C13 and C14

DESCRIPTION: West tower of 4 stages has stepped plinth, 3 string-courses, coved eaves with 4 spouts, crenellated parapet. To west, 2 clasp buttresses. Second stage has, to north, south and west, a small lancet. Third stage has, to north and south, a round-headed opening with hood-mould; to east, doorway; to west, C13 triple lancet with hood-mould. Fourth stage has 4 C14 cusped double lancets with hood-moulds.

GRADE: I

HERITAGE LIST NUMBER: 1264623

338 Laneham St Peter

LOCATION: Main Street, Laneham

POSTCODE: DN22 0NQ

OS GRID REF: SK815765

DATE: C12

DESCRIPTION: C12 tower with C15 diagonal buttresses and battlements. West doorway with double-chamfered pointed arch and continuous hood-mould. Above is an arched 2-light Y-traceried window and continuous hood-mould with carved human head at the apex. The 4 bell-chamber openings each consist of a pair of arched lights with hood-mould and label stops. There is a small carved human head to the centre of the western opening.

GRADE: I

HERITAGE LIST NUMBER: 1233511

339 Langford St Bartholomew

LOCATION: Holme Lane, Langford

POSTCODE: NG23 7RT

OS GRID REF: SK821590

DATE: C13

DESCRIPTION: The C13 tower of 2 stages is rendered with ashlar quoins and dressings. The second floor south has a small lancet window, with a larger one to the first floor west. The bell-chamber has four 2-light traceried openings with hood-mould and label stops, the eastern one having a centre carved head to the hood-mould. There are 4 carved heads to the corners of the tower which is embattled with a crocketed pinnacle to each corner.

GRADE: I

HERITAGE LIST NUMBER: 1302343

340 Mattersey All Saints

LOCATION: Abbey Road, Mattersey

POSTCODE: DN10 5DX

OS GRID REF: SK691893

DATE: C15

DESCRIPTION: The diagonal-buttressed tower of 2 stages has a restored, arched 3-light window with reticulated tracery, cusping, hood-mould and label stops. The belfry has 4 arched 2-light openings with panel tracery and cusping, with hood-mould and label stops over. Under, on the south and west sides, are single early C20 clock faces. There are 2 gargoyles to all but the west wall, which has a single remaining one. The parapet is embattled with single crocketed pinnacles at the angle.

GRADE: I

HERITAGE LIST NUMBER: 1239190

341 Misson
St John the Baptist

LOCATION: Church Street, Misson

POSTCODE: DN10 6EG

OS GRID REF: SK691949

DATE: C14 and C19

DESCRIPTION: West tower of 2 stages; moulded plinth, 2 string-courses, pair of gargoyles to north and south, crenellated parapet with 4 crocketed pinnacles. To west, pair of angle buttresses, 3 set-offs; pair of buttresses to east, engaged with aisles, 2 set-offs. To west, C14, chamfered and rebated doorway with hood-mould containing close-boarded door. Above, C14 chamfered and rebated opening with hood-mould, containing cusped-head triple lancet with panel tracery. Above again, clock dial. To south above, single unglazed lancet. Second stage has 4 C19 chamfered and rebated openings with label moulds, containing cusped ogee-head double lancet.

GRADE: I

HERITAGE LIST NUMBER: 1045108

342 Normanton on Trent
St Matthew

LOCATION: Tuxford Road, Normanton on Trent

POSTCODE: NG23 6RN

OS GRID REF: SK790690

DATE: Early C15

DESCRIPTION: The diagonally buttressed tower with C20 parapet is set on a plinth with heavy moulded band over and string-course at each juncture of the 3 stages. There is a single gargoyle to each side. The arched west window has 3 cusped lights. In the belfry are 4 C15 arched 2-light openings with panel tracery. There are 2 rectangular stair lights and 4 round tie-plates. On the east is a single clock face.

GRADE: II*

HERITAGE LIST NUMBER: 1233792

343 North Muskham St Wilfrid

LOCATION: Main Street, North Muskham

POSTCODE: NG23 6HD

OS GRID REF: SK798586

DATE: C14 and C15

DESCRIPTION: West tower of 2 stages, has chamfered plinth, string-course, eaves band with masks and 8 gargoyles, moulded crenellated parapet. 4 corner buttresses, single set-off, those to east engaged with aisles. First stage has, to west, a cusped ogee-headed lancet in square-headed reveal with hood-mould, and above, a single lancet. Above again, clock. Second stage has 4 C15 cusped-head double-lancet bell-openings with Decorated tracery, chamfered mullions, chamfered and rebated reveals with hood-moulds, mask and shield stops. All restored.

GRADE: I

HERITAGE LIST NUMBER: 1178997

344 North Wheatley St Peter & St Paul

LOCATION: Church Street, North Wheatley

POSTCODE: DN22 9BZ

OS GRID REF: SK762859

DATE: c.1480

DESCRIPTION: West tower, 2 stages, has string-course, eaves band with remains of gargoyles, crenellated parapet. 2 corner buttresses, 3 set-offs. To west, C19 cusped-head triple lancet with panel tracery and splayed reveal with hood-mould. Second stage has 4 double-lancet bell-openings with cusped trefoil heads.

GRADE: II*

HERITAGE LIST NUMBER: 1234572

345 Norton Cuckney St Mary

LOCATION: Norton Lane, Norton Cuckney

POSTCODE: NG20 9JR

OS GRID REF: SK567714

DATE: C12–C16

DESCRIPTION: The embattled tower with 4 crocketed pinnacles and 7 worn gargoyles is set on a plinth with a heavy moulded band over. There is a string-course at the juncture of the 2 stages, the top stage being ashlar. The first C12 stage with ashlar quoins has a clasping stair-turret at the south-west corner with single rectangular lights and single round tie-plates to the west and south. The arched west doorway with inner chamfered arch and wooden door has a hood-mould. Above, under a flat head, is a C16 2-light window with single stone mullion, probably replacing an arched opening. There is a single round tie-plate with 2 similar tie-plates on the north side. Above, to south and west, are single C20 clock faces. In the bell-chamber are 4 arched, mid-C13, double 2-light openings with hood-moulds. The south and east have single decorative shaped openings in the spandrel. Over all the bell-chamber openings are single cross tie-plates.

GRADE: I

HERITAGE LIST NUMBER: 1206551

346 Norwell St Laurence

LOCATION: Main Street, Norwell

POSTCODE: NG23 6JT

OS GRID REF: SK776617

DATE: C13 with C15 and C18 additions

DESCRIPTION: West tower of 4 stages, 4 string-courses, eaves band, 4 gargoyles, crenellated parapet with 4 crocketed pinnacles. Pyramidal roof with weathervane. To west, 2 diagonal buttresses, 4 set-offs, that to south-west inscribed "Gervis Downing & Samuell Curtiss Churchwardens 1713". First stage has, to south, a single lancet. To west, restored doorway with hood-mould and above, single lancet; to its left, 2 stair lights. Second stage has, to south, single lancet. To west, single lancet and stair light; above, clock. To north, single lancet. Third stage has 4 paired C13 lancets with chamfered mullions. Fourth stage has 4 C15 triple lancets in splayed and rebated reveals with 4-centred arched heads.

GRADE: I

HERITAGE LIST NUMBER: 1369970

347 Ordsall
All Hallows

LOCATION: All Hallows Street, Ordsall

POSTCODE: DN22 7TP

OS GRID REF: SK704797

DATE: C15

DESCRIPTION: The tower is short, of 3 stages with diagonal buttresses. It is perhaps C15, though much refaced. The upper parts were repaired after lightning damage in 1823, possibly by Richard Ingleman (who designed Ordsall Rectory in 1819 and was working at St Michael, West Retford, c.1822–24) and further renewed in 1877. The bell-openings are of 2 lights, with Perpendicular tracery. There is a plain west door with a 3-light window above.

GRADE: II*

HERITAGE LIST NUMBER: 1045228

348 Oxton
St Peter & St Paul

LOCATION: Main Street, Oxton

POSTCODE: NG25 0SA

OS GRID REF: SK630514

DATE: Late C14

DESCRIPTION: Diagonally buttressed tower of 2 stages with bands, set on a plinth with damaged moulded band over and embattled with remains of single corner pinnacles. The west side has a single-arched C14 window with 3 arched and cusped lights, reticulated tracery and hood-mould. The 4 C14 bell-chamber openings each have 2 arched and cusped lights and cusped tracery. Under the north and south openings are single clock faces. The south and north sides have single rectangular lights.

GRADE: I

HERITAGE LIST NUMBER: 1285622

349 Rampton
All Saints

LOCATION: Torksey Street, Rampton

POSTCODE: DN22 0HT

OS GRID REF: SK799786

DATE: Early C13 with C14 additions

DESCRIPTION: Early C13 tower of 2 stages with bands and course of ball-flower running under the C14 battlements. West moulded pointed-arch doorway with hood-mould and label stops. Above is an arched 3-light late C14 window with panel tracery, cusping, moulded surround, hood-mould and worn beast label stops, flanked by single corbels. Above is an arched and cusped niche with a moulded arched surround with finial. In the belfry are 4 arched c.1300 openings each with 2 cusped lights and hood-mould. There is a single clock face to the south and 5 small rectangular lights to the west, with single similar lights to the north and east.

GRADE: I

HERITAGE LIST NUMBER: 1233879

350 Rolleston Holy Trinity

LOCATION: Station Road, Rolleston

POSTCODE: NG23 5SE

OS GRID REF: SK742524

DATE: C12 and C14

DESCRIPTION: Angle-buttressed tower set on a moulded plinth and of 4 stages with bands. Embattled with 8 crocketed pinnacles, the centre pinnacle on each side being corbelled out, the corbel rising from the centre point between the 2-light bell-chamber openings. There are 8 gargoyles. The west wall has a single ogee-arched light originally C12, recut C14. Above is an arched 2-light window with ogee-arched and cusped lights and single quatrefoil. The north and south sides have single, similar recut lights. There are 4 C14 bell-chamber openings, each with 2 pointed-arch openings surmounted by a further 2 similar openings. The ogee-arched hood-mould is broken by the pinnacle corbel. The south side has a single rectangular stair light and the south and north sides have single small arched and cusped lights and 4 tie-plates. The west side has 2 tie-plates and on each side is a single large tie-plate.

GRADE: I

HERITAGE LIST NUMBER: 1045559

351 Saundby St Martin

LOCATION: Gainsborough Road, Saundby

POSTCODE: DN22 9ER

OS GRID REF: SK785880

DATE: c.1504

DESCRIPTION: West tower, ashlar, of 3 stages on plinth, 2 string-courses, battlemented parapet, 8 crocketed pinnacles, 2 diagonal buttresses to west, 3 set-offs; plain buttresses to north-east and south-east, 4 set-offs that to north engaged with aisle. West window C15 cove-moulded surround containing 3 round-headed lights with chamfered mullions and 4 similar lights above. Third stage has 4 double-lancet bell-openings with cusped heads; C19 iron clock dial to west.

GRADE: I

HERITAGE LIST NUMBER: 1045083

352 Screveton St Wilfred

LOCATION: Car Colston Road, Screveton

POSTCODE: NG13 8JT

OS GRID REF: SK729434

DATE: c.1475

DESCRIPTION: West tower of 3 stages, has 2 coved string-courses, eaves band, crenellated parapet, 4 pinnacles. Pair of diagonal buttresses to west, 3 set-offs. Pair of buttresses to east, engaged with aisles. To west, C16 triple lancet in chamfered square-headed reveal with hood-mould. Above, slatted segmental-headed opening. Above again, each side has a round-headed double lancet with hood-mould.

GRADE: I

HERITAGE LIST NUMBER: 1243811

353 Selston St Helen

LOCATION: Church Lane, Selston

POSTCODE: NG16 6EW

OS GRID REF: SK458533

DATE: Late C14

DESCRIPTION: West tower of 2 stages, has chamfered and moulded plinth, string-course, eaves band with 4 mask spouts, crenellated parapet with inscription and arms of Beauvale Priory. 4 diagonal buttresses, 6 set-offs, those to east merging with aisles. First stage has, to north and south, a single light. To west, chamfered doorway with hood-mould; above it, cusped triple lancet with panel tracery and hood-mould. Second stage has, on each side, double lancet with Y-tracery and hood-mould.

GRADE: II*

HERITAGE LIST NUMBER: 1275170

354 Shelford St Peter & St Paul

LOCATION: Church Street, Shelford

POSTCODE: NG12 1EJ

OS GRID REF: SK661424

DATE: C15

DESCRIPTION: Angle-buttressed embattled tower set on a moulded plinth and of 3 stages. Battlements with some blind tracery. West side with moulded arched doorway and double C19 door. Above is a single restored arched 3-light window with panel tracery, hood-mould and label stops. There are 3 rectangular stair lights. Projecting from the south side is a tall narrow ashlar lean-to. There is a single stair light. The east side has a single stair light. The 4 C15 arched bell-chamber openings each have 3 cinquefoil cusped lights with hood-mould and label stops. Under the south and east openings are single clock faces.

GRADE: II*

HERITAGE LIST NUMBER: 1250021

355 Sibthorpe St Peter

LOCATION: Church Lane, Sibthorpe

POSTCODE: NG23 5PT

OS GRID REF: SK763454

DATE: C13

DESCRIPTION: West tower of 2 stages, has to west 2 pairs of corner buttresses, 3 set-offs, with chamfered plinths. String-course, moulded eaves, plain parapet. First stage has to south, chamfered lancet. To west, restored C14 doorway with moulded imposts, hood-mould and mask stops; above, rectangular light. East side has remains of 2 nave gables. Second stage has 4 paired lancets with central shaft, vesica piscis and hood-mould.

GRADE: I

HERITAGE LIST NUMBER: 1244670

356 Skegby St Andrew

LOCATION: Mansfield Road, Skegby

POSTCODE: NG17 3DD

OS GRID REF: SK492610

DATE: C13

DESCRIPTION: Unbuttressed west tower of 2 stages, has chamfered string-course, eaves band and coped parapet with remains of 8 pinnacles. First stage has round-headed window to south and to west, C19 double lancet with Y-tracery and hood-mould. Second stage has 4 round-headed double openings.

GRADE: II

HERITAGE LIST NUMBER: 1234873

357 South Leverton All Saints

LOCATION: Church Street, South Leverton

POSTCODE: DN22 0BX

OS GRID REF: SK783811

DATE: C12 and C15

DESCRIPTION: West tower of 3 stages, has chamfered plinth, 2 string-courses, eaves band, crenellated parapet, remains of partial buttress to south-east. North, south and east sides have each a single gargoyle. West side has bolection-moulded square-headed blocked doorway, and above, C14 single lancet with splayed reveal and hood-mould. Above again, single round-headed light and clock, 1977. Second stage has 4 round-headed openings with colonnettes, each containing 2 round-headed bell-openings with central shaft and square capitals.

GRADE: II*

HERITAGE LIST NUMBER: 1275880

358 South Muskham St Wilfrid

LOCATION: Church Lane, South Muskham

POSTCODE: NG23 6EQ

OS GRID REF: SK794573

DATE: C13 and C15

DESCRIPTION: West tower of 4 stages. Deep-moulded plinth, 3 string-courses, coved eaves band with 4 gargoyles and spouts to north and south, all restored. Crenellated parapet with 4 truncated pinnacles. Weathercock. 2 corner buttresses to west, becoming pilasters above second stage, 6 set-offs. 2 plain buttresses to east, 6 set-offs, engaged with aisles. First stage has, to south, a C14 trefoil-headed lancet and, to west, chamfered C13 lancet. Second stage has, to south and west, a C14 double lancet with cusped head, chamfered mullions, chamfered and rebated reveal, and hood-mould. Third stage has, to west, clock. Fourth stage has 4 C15 double-lancet bell-openings with cusped heads and billeted transoms, set in double-lancet openings with chamfered reveals, coved ogee hood-moulds with finials and mask stops. To west, figure in hood-mould spandrel.

GRADE: I

HERITAGE LIST NUMBER: 1179422

359 South Scarle St Helena

LOCATION: Main Street, South Scarle

POSTCODE: NG23 7JX

OS GRID REF: SK848640

DATE: C14

DESCRIPTION: West tower of 4 stages, with battlemented ashlar parapet and 8 crocketed pinnacles; clasping buttresses, 4 set-offs, on plinths, up to third stage. String-courses at each stage. Triple lancet on second stage, with intersecting tracery and hood-mould. Bell stage has 4 double lancets. South porch, C15, coursed rubble with ashlar dressings, square plan, with 2 diagonal buttresses with 3 set-offs; plinth, moulded string-course, upper string-course, battlemented parapet with 2 humanistic gargoyles. 4-centred arched doorway with rebate, roll and cove mouldings; respond shafts with crenellated octagon capitals. Above, niche containing female figure.

GRADE: I

HERITAGE LIST NUMBER: 1302469

360 Stanford on Soar St John the Baptist

LOCATION: Main Street, Stanford on Soar

POSTCODE: LE12 5PY

OS GRID REF: SK543220

DATE: C14

DESCRIPTION: Diagonal-buttressed tower of 3 stages with bands, on a moulded plinth and with single corner gargoyles. Under the battlements extends a frieze of quatrefoils set into lozenges. The west side has a moulded arched doorway with Tudor hood-mould and label stops, the spandrels decorated with single trefoils containing blind shields. Over is a single C14 arched 3-light window with cusped tracery, hood-mould and figurative label stops. Above is a single C14 2-light window with cusped tracery under a flat arch with hood-mould and human head label stops. There are 3 stair lights. 4 arched bell-chamber openings each with 2 cinquefoil-arched lights, hood-mould and human head label stops, under those on the north, east and south sides are single clock faces.

GRADE: 1

HERITAGE LIST NUMBER: 1242187

361 Strelley
All Saints

LOCATION: Main Street, Strelley

POSTCODE: NG8 6PE

OS GRID REF: SK506421

DATE: C13

DESCRIPTION: West tower of 3 stages, has a small central buttress on each side. 2 string-courses. Pyramid roof with weathercock. First stage has, to south, 2 chamfered lancets, the upper one smaller; to west, similar lancet and stair light. Second stage has, on each side, remains of blocked C13 double lancet with trefoil head and hood-mould; to south, clock. Third stage has, on each side, double-lancet opening with Tudor-arched head and hood-mould.

GRADE: I

HERITAGE LIST NUMBER: 1248224

362 Sturton le Steeple St Peter & St Paul

LOCATION: Church Street, Sturton le Steeple

POSTCODE: DN22 9HQ

OS GRID REF: SK788838

DATE: Mid-C14 and C15

DESCRIPTION: The west tower has a moulded and chamfered plinth, 3 string-courses, eaves band, 8 gargoyles, crenellated parapet, 12 crocketed pinnacles. To west, 2 corner buttresses, 3 set-offs, and intermediate buttress, single set-off. To east, 2 buttresses engaged with aisles, 3 set-offs. First stage has, to south, 4-centred arched door and above, a trefoil-headed lancet. To west, a cusped-head double lancet with Decorated tracery, splayed mullions and chamfered and rebated reveal; above, 2 stair lights and trefoil-headed lancet. North side has a single similar lancet. Second stage has on each side a C14 cusped-head double lancet. Third stage has a clock face to west and to north. Fourth stage has a C15 transomed triple-lancet bell-opening with panel tracery on each side.

GRADE: II*

HERITAGE LIST NUMBER: 1275773

363 Sutton cum Lound St Bartholomew

LOCATION: Mattersey Road, Sutton cum Lound

POSTCODE: DN22 8PL

OS GRID REF: SK681850

DATE: C15

DESCRIPTION: The angle-buttressed tower of base and 2 stages has a C15 arched 3-light west window with panel tracery. There is a hood-mould over and a string-course under. The bell-chamber has 4 C15 2-light arched openings each with panel tracery, cusping, hood-mould and label stops. The tower is embattled with 8 crocketed pinnacles. Guttering runs under, having 2 gargoyles to each side. Under the south and east bell-chamber openings are single C20 clock faces. There is a single small rectangular light to the south wall, with 3 similar to the west.

GRADE: I

HERITAGE LIST NUMBER: 1239776

364 Syerston All Saints

LOCATION: Syerston

POSTCODE: NG23 5NE

OS GRID REF: SK746474

DATE: C13

DESCRIPTION: C13 low single-stage tower, embattled C19, with single lancet on the south side and single rectangular stair light above. The 4 arched bell-chamber openings each have 2 pointed-arch lights, hood-mould and remains of label stops.

GRADE: II

HERITAGE LIST NUMBER: 1178850

365 Thurgarton St Peter

LOCATION: Priory Road, Thurgarton

POSTCODE: NG14 7GX

OS GRID REF: SK692491

DATE: C13

DESCRIPTION: The embattled tower with remnants of 4 pinnacles is of 6 stages with bands. The clasping buttresses rise to the bell-chamber openings, the left buttress being larger. West wall with pointed-arch doorway originally with 5 colonnettes either side, but the left now with 3 and the right with 2, alternating a larger with smaller. The moulded arch has 3 bands of dog-tooth with a dog-tooth hood-mould and right impost band. Above is a damaged 4-bay blind arcade with pointed chamfered arches, colonnettes and continuous hood-mould. Above is a pointed-arch window with dog-tooth hood-mould and worn label stops. In the left buttress is a 3-bay blind arcade with colonnettes, moulded pointed arches and continuous hood-mould with 3 worn label stops, the central arcade with single rectangular light. The right buttress has a single similar larger arcade with 2 engaged octagonal columns and hood-mould. Extending over all and stepped to accommodate the central window is a continuous hood-mould. Above this window is a single rectangular light and in the right buttress a single blind arcade with 2 colonnettes either side supporting a moulded pointed arch, containing a single rectangular light. Above, traversing tower and buttresses, is a 6-bay blind arcade with pointed moulded arched colonnettes, some with nail-head capitals and continuous nail-head hood-mould. The single arcade of the right buttress has a single carved grotesque. The tower with 3 arcades, the outer each with a single lancet with chevron decoration to the arch. Over the right arcade of the left buttress and the arcade of the right buttress are triangular arches decorated with dog-tooth, marking the termination of the buttresses. At the angles of each buttress and at stages where there is no decoration are single engaged colonnettes. The north side of the tower with similar clasping buttresses, that on the right being larger, is set on a plinth with moulded band over. At the first stage is a single rectangular stair light. Above, in the right buttress, is a 3-bay blind arcade with pointed moulded arches, the central arch decorated with dog-tooth, colonnettes and hood-mould with 4 decorative label stops. The east side of the buttress has a single similar arcade. Above, in the tower, is a single lancet and, in the right buttress, east side, a single rectangular stair light. Above, traversing tower and buttresses, is a 6-bay blind arcade corresponding to that at the same stage on the west side with 2 similar lights in the tower arcading. However, all but 2 of the capitals are decorated with worn nail head. There are 3 decorative label stops and the right arcade of the right buttress has a single small arched light with dog-tooth hood-mould. Above, in the buttresses, are triangular arches with dog-tooth hood-moulds, the right arch with single small arched light with chevron to jambs and arch. The east side has a single lancet. Above, traversing this front, is a 4-bay blind arcade with pointed moulded arches and colonnettes with nail-head capitals and continuous nail-head hood-mould with 3 label stops. The left and third-left arcades have single lancets with chevron decoration to the arches. The second-left arcade has a single clock face. The south front has 2 arched lights with continuous nail-head hood-mould. Each side of the bell-chamber has 2 arched openings, each opening flanked by single colonnettes with moulded capitals which rise from impost level. Either side are 2-bay blind arcades with similar colonnettes and pointed moulded arches. Over, extending around all 4 sides, is a continuous nail-head hood-mould with 20 decorative label stops. Attached to the north-west buttress is a dressed coursed rubble wall with ashlar coping having a pointed-arch doorway with hood-mould and 2 large grotesque corbels to the west side and a single carved human head to the apex of the east side.

GRADE: I

HERITAGE LIST NUMBER: 1045537

366 Treswell St John the Baptist

LOCATION: New Road, Treswell

POSTCODE: DN22 0EQ

OS GRID REF: SK781793

DATE: C14 and C15

DESCRIPTION: Diagonal-buttressed tower of 2 stages with string-course at the juncture. Single gargoyle to each side. Single-arched window with 3 arched lights, string-course under and hood-mould. There are 4 arched bell-chamber openings, each with 2 arched lights, tracery, cusping and hood-mould, and 4 rectangular stair lights.

GRADE: I

HERITAGE LIST NUMBER: 1234072

367 Trowell St Helen

LOCATION: Nottingham Road/Stapleford Road, Trowell

POSTCODE: NG9 3QW

OS GRID REF: SK484397

DATE: C13 and C15

DESCRIPTION: West tower of 2 stages, has moulded string-course and eaves, crenellated parapet and 4 corner buttresses. First stage has, to south, door, and to west, transomed triple lancet, C15, with panel tracery and hood-mould. Second stage has on each side a rectangular opening and above it a cusped transomed double lancet with hood-mould and lintel band. To north, clock removed from Nottingham Exchange.

GRADE: II*

HERITAGE LIST NUMBER: 1278008

368 Upper Broughton St Luke

LOCATION: Rectory Road, Upper Broughton

POSTCODE: LE14 3BG

OS GRID REF: SK683262

DATE: C13 and C14

DESCRIPTION: The 2-stage tower has later clasping buttresses to lower stage, small lancet west window, 2-light, louvred bell-chamber openings and quatrefoil frieze under crenellated parapet.

GRADE: I

HERITAGE LIST NUMBER: 1264583

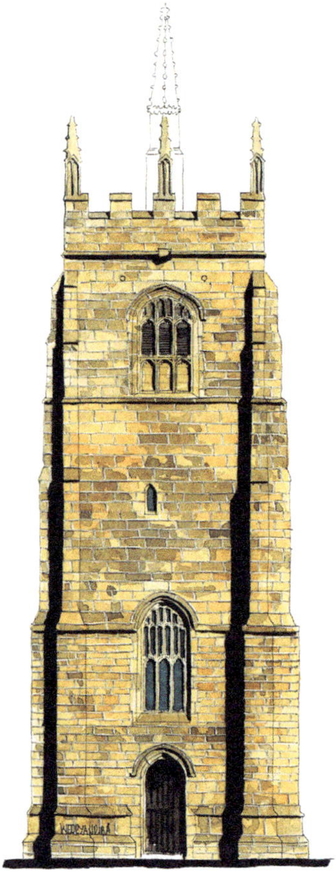

369 Upton St Peter & St Paul

LOCATION: Church Lane, Upton

POSTCODE: NG23 5SR

OS GRID REF: SK736542

DATE: C15

DESCRIPTION: West tower of 3 stages has deep moulded plinth, 2 string-courses, gargoyles to north, south and east, crenellated parapet, C15, with 8 crocketed pinnacles and single central pinnacle with weathercock. 4 corner buttresses, 4 set-offs plus plinths. First stage has, to south, 2 stair lights and chamfered unglazed opening. West side has chamfered and moulded doorway with 4-centred arched head and hood-mould, and C19 traceried panelled door. Above, C15 cusped-head triple lancet with panel tracery and moulded reveal with hood-mould. Second stage has, to south, a chamfered round-headed opening with grill; to north, a reset corbel. Third stage has 4 C15 cusped, transomed triple lancets with panel tracery, louvres and hood-moulds.

GRADE: I

HERITAGE LIST NUMBER: 1369997

370 Walesby St Edmund

LOCATION: Tuxford Road, Walesby

POSTCODE: NG22 9NY

OS GRID REF: SK685707

DATE: C15 and early C16

DESCRIPTION: West tower of 3 stages. 3 coved string-courses, 3 gargoyles and spout, crenellated parapet with Stanhope Arms in panel with hood-mould, cusped ogee cutout, and 4 crocketed pinnacles. 2 diagonal buttresses to west, 3 set-offs. First stage to west adjoining C20 boiler house. Above, late C15 triple lancet with ogee and round heads, panel tracery and hood-mould. Third stage has 4 louvred double lancets with 4-centred arched reveals and hood-moulds.

GRADE: II*

HERITAGE LIST NUMBER: 1178711

371 Walkeringham St Mary Magdalene

LOCATION: Gringley Road, Walkeringham

POSTCODE: DN10 4HZ

OS GRID REF: SK771922

DATE: C14 and C15

DESCRIPTION: West tower of 2 stages, has moulded plinth, 2 string-courses, crenellated parapet with 8 crocketed pinnacles; 2 diagonal buttresses, 3 set-offs; plain buttresses to north-east and south-east, 2 set-offs; single additional buttress to north-east, single set-off. West window C14, double coved surround with hood-mould, containing restored triple lancet with cusped heads, panel tracery and coved mullions. Above, clock faces to north and west. First stage has 4 C15 bell-openings with hood-moulds, containing triple lancets with panel tracery.

GRADE: I

HERITAGE LIST NUMBER: 1045085

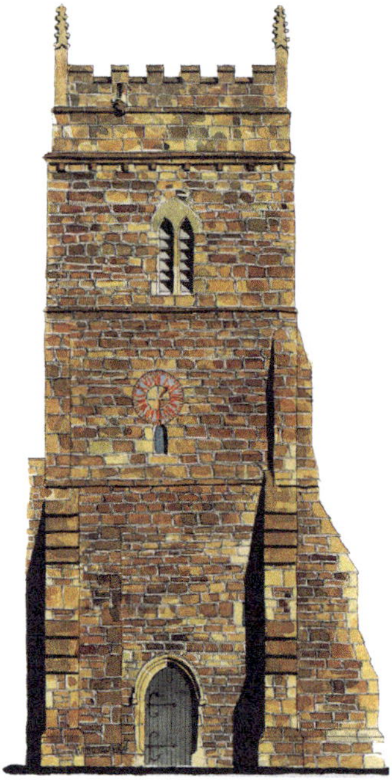

372 Wellow
St Swithin

LOCATION: Newark Road, Wellow

POSTCODE: NG22 0EA

OS GRID REF: SK671661

DATE: C12–C15

DESCRIPTION: Tower of 3 stages, has 2 string-courses, dentilled top band, coved eaves with an off-centre gargoyle on each side, crenellated parapet with 4 crocketed pinnacles. 4 buttresses, 3 set-offs, those to east engaged with aisle and nave. First stage has, to south, square blank recess; to west, restored C13 door with hood-mould. Second stage has, to west, chamfered C13 lancet and clock. Third stage has 4 plain chamfered louvred double lancets.

GRADE: II*

HERITAGE LIST NUMBER: 1370185

373 West Bridgford
St Giles

LOCATION: Bridgford Road, West Bridgford

POSTCODE: NG2 6AX

OS GRID REF: SK586375

DATE: C15

DESCRIPTION: West tower of single stage has moulded eaves with 2 gargoyles to north and south. To west, 2 diagonal buttresses. South side has, above, inscribed stone. West side has restored triple lancet with panel tracery. North-west corner has doorway. Above, each side has a double lancet in round-headed reveal.

GRADE: II

HERITAGE LIST NUMBER: 1045677

374 Widmerpool St Peter & St Paul

LOCATION: Old Hall Drive, Widmerpool

POSTCODE: NG12 5PZ

OS GRID REF: SK628282

DATE: C14

DESCRIPTION: Single-stage angle-buttressed embattled tower. Angles of battlements with evidence of single former pinnacles. Single gargoyle to each side. West side with attached stair-turret with single light. To the right is a C14 arched 3-light window with flowing tracery. Above are 2 rectangular lights. The north and south sides have single rectangular lights. Four 2-light C14 bell-chamber openings with mouchettes under flat arches.

GRADE: II*

HERITAGE LIST NUMBER: 1242476

375 Winkburn St John of Jerusalem

LOCATION: Church Lane, Winkburn

POSTCODE: NG22 8PQ

OS GRID REF: SK711583

DATE: Mid-C12

DESCRIPTION: Diagonally buttressed embattled tower with single C19 crocketed pinnacles at the angles. Set on a plinth, chamfered to the south, of 3 stages with bands. West side has a single-arched C14 window with 2 arched and cusped lights. Above is a single blocked lozenge-shaped opening. The north side has a single rectangular light and above a single recessed panel with decoratively carved surround inscribed "New Built Ano Domi 16.2 William B … Thomas W …." The south side has 2 rectangular lights. The 4 C12 arched bell-chamber openings are decorated with chevrons to jambs and arch on the west side, the north side with chevron decoration remaining to jambs only. The south side has cable moulding to jambs and arch, and the east side cable moulding to arch, chevrons to jambs and inner arches. The inner arches of the other 3 openings with worn decoration. Over the south opening is a single rectangular blind panel, and below the east opening is a raised and stepped band.

GRADE: I

HERITAGE LIST NUMBER: 1285594

376 Woodborough St Swithun

LOCATION: Main Street, Woodborough

POSTCODE: NG14 6DX

OS GRID REF: SK632477

DATE: C13 and C15

DESCRIPTION: West tower of 2 stages, has coved string-course and eaves band, crenellated parapet and 4 crocketed pinnacles. Pair of diagonal buttresses, 3 set-offs. First stage has, to south, a light. West side has chamfered 4-centred arched doorway and, above it, similar C15 triple lancet. North side has clock. Second stage has 4 restored cusped double lancets.

GRADE: II*

HERITAGE LIST NUMBER: 1264188

Diocese of

York

377 Aldbrough St Bartholomew

LOCATION: Church Street, Aldbrough

POSTCODE: HU11 4RN

OS GRID REF: TA244387

DATE: Early C13

DESCRIPTION: 3-stage west tower. Moulded plinth, angle buttresses, side-alternate quoins, scroll-moulded strings between stages. Lancets to first and second stage. Third stage north elevation has unaltered belfry openings of 2 pointed lights with mid-wall shaft under round roll-moulded arch on nook-shafts. Remaining elevations have pointed 3-light windows with Perpendicular tracery. Dentilled brick parapet.

GRADE: II*

HERITAGE LIST NUMBER: 1083529

378 Allerston St John

LOCATION: Church Lane, just off A170, Allerston

POSTCODE: YO18 7PP

OS GRID REF: SE878829

DATE: C15

DESCRIPTION: 3-stage tower with diagonal buttresses and recessed pointed door beneath hood-mould to western elevation. 3-light window with panel tracery and stopped pointed hood-mould above, the stops carved as shields bearing traces of the emblems of the Knights Hospitaller and the Hastings family. 2-light transomed bell-openings with louvres beneath stopped pointed hood-moulds to each face. Parapet, carved with a band of blind tracery, is supported by a corbel-table carved with a variety of motifs.

GRADE: II*

HERITAGE LIST NUMBER: 1149549

379 Appleton-le-Street All Saints

LOCATION: Main Road, Appleton-le-Street

POSTCODE: YO17 6PG

OS GRID REF: SE734735

DATE: C11

DESCRIPTION: Tower of 3 stages with original west doorway now blocked by round-arched light with stopped hood-mould.

GRADE: I

HERITAGE LIST NUMBER: 1296551

380 Aughton All Saints

LOCATION: Track at the end of Main Street, Aughton

POSTCODE: YO42 4PQ

OS GRID REF: SE702387

DATE: Early C16

DESCRIPTION: Perpendicular 3-stage tower with diagonal stepped buttresses carrying grotesque figures. Plain 5-light west window. South side: row of relief shields and illegible inscription dated 1536 with 3-light square-headed window beneath hood-mould above. 2-light square-headed belfry openings to each face except triple trefoil-headed lancet window to east. Embattled parapet with 4 pinnacles.

GRADE: I

HERITAGE LIST NUMBER: 1346742

381 Bainton St Andrew

LOCATION: St Andrew's Close, Bainton

POSTCODE: YO25 9NL

OS GRID REF: SE965523

DATE: C14

DESCRIPTION: Tower has a chamfered plinth and angle buttress with offsets to full height to western corners. Stair-turret to south-west corner. To west a 4-light, pointed window with reticulated tracery to head under hood-mould with face stops in double-chamfered surround. First-stage band. To west side a slit window. Second-stage band. 2-light bell-openings with pointed heads and reticulated tracery with hood-moulds with face stops to each side. Corbel-table. Battlements pierced with crosses.

GRADE: I

HERITAGE LIST NUMBER: 1083830

382 Beeford St Leonard

LOCATION: Rectory Lane, Beeford

POSTCODE: YO25 8BA

OS GRID REF: TA129543

DATE: C15

DESCRIPTION: West tower of 3 stages with chamfered plinth and string-courses and pinnacled diagonal buttresses. Quatrefoil window to first floor. 2-light belfry openings with transoms and Perpendicular tracery under pointed heads with hood-moulds. Pierced parapet with 8 diagonally set crocketed pinnacles. 3-light west window with brattished transoms and Perpendicular tracery under pointed head with hood-mould with grotesque stops. Niche above with figure of St Leonard on raised base under nodding ogee canopy with foliated finial.

GRADE: II*

HERITAGE LIST NUMBER: 1083372

383 Bempton St Michael

LOCATION: Church Lane, Bempton

POSTCODE: YO15 1HJ

OS GRID REF: TA191720

DATE: C14

DESCRIPTION: West tower of 2 plain lower stages and octagonal belfry stage, separated by string-courses. 2-light belfry openings with curvilinear tracery under battlemented parapet. West window of 2 trefoil-headed lights with quatrefoil over.

GRADE: II*

HERITAGE LIST NUMBER: 1161842

384 Birkin St Mary

LOCATION: Main Street, Birkin

POSTCODE: WF11 9LR

OS GRID REF: SE530265

DATE: C12 and C14

DESCRIPTION: Tower of 2 stages. Slit openings to lower stage. Band. C14 upper stage. 2-light belfry windows with reticulated tracery. Band with gargoyles. Battlements and pinnacles.

GRADE: I

HERITAGE LIST NUMBER: 1316671

385 Bishop Burton All Saints

LOCATION: Church Side, Bishop Burton

POSTCODE: HU17 8QB

OS GRID REF: SE990397

DATE: Early C13

DESCRIPTION: West tower with chamfered plinth. South door of 2 pointed and chamfered orders, the inner on carved corbels under hood-mould. Chamfered imposts. Chamfered string between stages. Paired lancets under round arch to belfry. Low parapet with moulded coping. West wall: projecting carved head to bottom stage, lancet to second stage, similar belfry openings. North side is identical to the south side with the addition of a clock face to the second stage.

GRADE: II*

HERITAGE LIST NUMBER: 1103429

386 Bolton Percy All Saints

LOCATION: Main Street, Bolton Percy

POSTCODE: YO23 7AL

OS GRID REF: SE531412

DATE: C14–C15

DESCRIPTION: 3-stage west tower; angle buttresses with offsets, chamfered plinth. West side has pointed-arch plank door under hood-mould; above, a 3-cinquefoil-light window under hood-mould. To second stage a 2-light straight-headed window within chamfered recess to south side. To third stage are 3-light cusped, straight-headed bell-openings. Battlements with angle pinnacles.

GRADE: I

HERITAGE LIST NUMBER: 1296630

387 Boynton St Andrew

LOCATION: Main Street, Boynton

POSTCODE: YO16 4XJ

OS GRID REF: TA136679

DATE: C15

DESCRIPTION: West tower of 3 stages with moulded plinth, diagonal buttresses, and string-courses. West window of 2 lights with Perpendicular tracery, altered to accommodate double west door with Gothic panelling. South wall has large sculpted figure of St Andrew in canopied niche. 2-light belfry opening with Perpendicular tracery under hood-mould, battlemented parapet with C18 pinnacles.

GRADE: I

HERITAGE LIST NUMBER: 1083392

388 Brafferton St Peter

LOCATION: School Lane, Brafferton

POSTCODE: YO61 2NZ

OS GRID REF: SE436701

DATE: C15

DESCRIPTION: Battlemented 3-stage tower with set-offs. 3-light window with C20 glazing, tiny trefoil window and 2-light bell-openings.

GRADE: II*

HERITAGE LIST NUMBER: 1314926

389 Brandesburton St Mary

LOCATION: Church Drive, Brandesburton

POSTCODE: YO25 8RG

OS GRID REF: TA119476

DATE: C13

DESCRIPTION: West tower of 3 stages with chamfered plinth and offsets and diagonal buttresses. C19 window with ogee tracery to second stage, paired lancets in square-headed opening to belfry. Crenellated parapet. West window of paired lancets in square-headed opening beneath ogee-headed niche.

GRADE: I

HERITAGE LIST NUMBER: 1263780

390 Brompton St Thomas

LOCATION: Church View, Brompton

POSTCODE: DL6 2QZ

OS GRID REF: SE373963

DATE: C15

DESCRIPTION: South-west tower of 3 stages, offset diagonal buttresses to first stage. Plinth. Board south door in chamfered basket-arched surround with hood-mould and small niche above. Projection for staircase to east side. Chamfered band, second stage, clock faces to east and west sides, chamfered band. Chamfered, cusped 2-light flat-headed belfry openings. Band, embattled parapet with crocketed pinnacles to each corner. Pyramidal roof with flagpole.

GRADE: I

HERITAGE LIST NUMBER: 1150903

391 Bubwith All Saints

LOCATION: Church Street, Bubwith

POSTCODE: YO8 6LS

OS GRID REF: SE711361

DATE: C15

DESCRIPTION: 2-stage tower with diagonal buttresses. Tall 3-light west window. Traceried 2-light pointed belfry opening with hood-mould resting on string-course to each face. Embattled parapet with 8 crocketed pinnacles. Tower and clerestory built with funds from the will of Nicholas de Bubwith, Bishop of Bath and Wells (1408–24), possibly designed by Robert Cooper (fl.1442–59).

GRADE: I

HERITAGE LIST NUMBER: 1310699

392 Bugthorpe St Andrew

LOCATION: Main Street, Bugthorpe

POSTCODE: YO41 1QL

OS GRID REF: SE772578

DATE: C15

DESCRIPTION: West tower, moulded plinth, moulded string to belfry stage; pointed 2-light belfry openings under 4-centred arch with continuous hollow chamfer. Gargoyles beneath crenellated parapet with crocketed corner pinnacles. Pointed west window of 2 lights with Perpendicular tracery under a hood-mould. In angle between north wall of tower and west wall of nave is a square stair-turret with slit windows, beneath a corbelled roof.

GRADE: I

HERITAGE LIST NUMBER: 1309966

393 Bulmer St Martin

LOCATION: Main Street, Bulmer

POSTCODE: YO60 7BN

OS GRID REF: SE699676

DATE: C15

DESCRIPTION: C15 tower with date stone on parapet of 1637. 2-light square-headed window flanked by diagonal buttresses to west front of 3-stage tower. Illegible memorial slab with Greek key border to south face. String-course to second stage carries small lancet window to west. Double-lancet belfry openings to each face. Later embattled parapet with pinnacles and date stone.

GRADE: I

HERITAGE LIST NUMBER: 1174317

394 Burstwick All Saints

LOCATION: Church Lane, Burstwick

POSTCODE: HU12 9ER

OS GRID REF: TA227280

DATE: Mid-C15

DESCRIPTION: Mid-C15 3-stage tower incorporating a mid-C13 west window. Moulded plinth, diagonal buttresses with offsets, weathered string-courses between stages, stair-lighting slits to south-west angle. First stage: pointed transomed 3-light west window with curvilinear tracery in hollow-chamfered reveal with hood-mould; trefoiled ogee-headed niche above with crocketed hood, carved corbel of angel holding shield, and flanking pinnacled buttress shafts. Second stage: slit lights to north and south, and outline of earlier nave gable to east. Third stage: pointed 2-light traceried belfry openings with hollow-chamfered reveals; moulded string-course, coped embattled parapet with bases of former angle pinnacles, and carved head corbels to each side supporting shafts for former central pinnacles.

GRADE: I

HERITAGE LIST NUMBER: 1366240

395 Burton Agnes St Martin

LOCATION: Shady Lane, Burton Agnes

POSTCODE: YO25 4NB

OS GRID REF: TA102632

DATE: C15

DESCRIPTION: 3-stage west tower; plinth, diagonal buttresses with offsets. To west side a pointed plank door in triple-chamfered cavetto- and ovolo-moulded surround. Above, a 3-light Perpendicular window with partly recut mullions under hood-mould with angel stops, on cavetto-moulded continuous sill band. To second stage a 2-light flat-headed opening under hood-mould, between canopied niches for statues. To third stage are 3-stepped cinquefoil-light bell-openings in 4-centred arched surrounds. Eaves band with gargoyles at diagonals, battlements.

GRADE: I

HERITAGE LIST NUMBER: 1083812

396 Burton Fleming St Cuthbert

LOCATION: Front Street, Burton Fleming

POSTCODE: YO25 3PS

OS GRID REF: TA083724

DATE: C13

DESCRIPTION: West tower of 4 stages, with hollow-chamfered plinth to west face returning for a short distance to north and south, and massive buttresses to north-east and south-east angles, the latter cut back. 2-light west window with Decorated tracery. 2 string-courses and an offset at belfry floor level. Lancets to belfry, enriched by balls, chevrons and cusping. Coped parapet conceals low-pitched roof.

GRADE: II*

HERITAGE LIST NUMBER: 1083394

397 Burton Pidsea St Peter

LOCATION: Church Street, Burton Pidsea

POSTCODE: HU12 9AU

OS GRID REF: TA251311

DATE: C15

DESCRIPTION: West tower, moulded plinth, diagonal buttresses with offsets, chamfered strings between stages. Pointed 4-light west window. Pointed 2-light belfry openings with Perpendicular tracery. Crenellated parapet.

GRADE: I

HERITAGE LIST NUMBER: 1083494

398 Carnaby St John the Baptist

LOCATION: Church Lane, Carnaby

POSTCODE: YO16 4UR

OS GRID REF: TA144655

DATE: C14–C15

DESCRIPTION: Magnesian limestone tower, double-chamfered plinth, full-height angle and diagonal buttresses with offsets; to west end, first stage a 2-light, pointed window with Perpendicular tracery to head, mostly recut, within double-chamfered surround and under hood-mould; first stage band; slit window to south side; 2-light traceried Perpendicular bell-openings, within double-chamfered surrounds and under hood-moulds; stairwell to north-west corner. Embattled parapets and corner pinnacles.

GRADE: II*

HERITAGE LIST NUMBER: 1083821

399 Catton All Saints

LOCATION: Church Lane, Low Catton

POSTCODE: YO41 1EA

OS GRID REF: SE704539

DATE: C15

DESCRIPTION: South-west tower: plinth with diagonal buttresses with offsets, slit window to second stage, 2-light pointed belfry openings with Perpendicular tracery under hood-mould, crenellated parapet with corner finials.

GRADE: I

HERITAGE LIST NUMBER: 1346407

400 Cawood
All Saints

LOCATION: Church End, Cawood

POSTCODE: YO8 3SN

OS GRID REF: SE577377

DATE: Early C15

DESCRIPTION: Early C15 bell-tower in 3 stages has polygonal angle buttresses surmounted by crockets to first stage and diagonal buttresses with set-offs above. Ogee-arched 3-light window to ground floor with restored Perpendicular tracery. First stage: crocketed nodding-ogee-headed niche containing C20 statue of the Blessed Virgin Mary on pedestal; to north and south, trefoiled lights under flat arches with drip-moulds. Recessed bell stage: 3-light ogee-arched bell-openings with restored Perpendicular tracery under drip-moulds with carved stops. String-course, buttresses, battlements and angle pinnacles.

GRADE: I

HERITAGE LIST NUMBER: 1316657

401 Cayton
St John the Baptist

LOCATION: Church Lane, Cayton

POSTCODE: YO11 3RP

OS GRID REF: TA056833

DATE: C15

DESCRIPTION: 3-stage, embattled tower with pinnacles and diagonal buttresses with offsets. 2-light, flat-topped bell-openings to upper stage; similar opening to lowest stage to west; slit opening to south to light tower staircase.

GRADE: I

HERITAGE LIST NUMBER: 1148133

402 Coxwold St Michael

LOCATION: Village Street, up the hill from centre of Coxwold

POSTCODE: YO61 4AD

OS GRID REF: SE533772

DATE: c.1430

DESCRIPTION: Octagonal west tower with staircase turret in south angle; angle buttresses with gablets and detached square shafts above, set diagonally and linked to the tower by short flying buttresses, 3-light west window with cinque-cusping and Perpendicular tracery; south and west vents to ringing chamber; belfry windows on each side have 2 lights, central transom and Perpendicular tracery; C18 clock on east side; between pinnacles of buttress shafts, delicate filigree tracery.

GRADE: I

HERITAGE LIST NUMBER: 1150758

403 Crambe St Michael

LOCATION: Town Green, Crambe

POSTCODE: YO60 7JR

OS GRID REF: SE734648

DATE: C15

DESCRIPTION: Buttressed tower of 3 stages divided by string-courses. Round-headed door in moulded arch with head stops to hood-mould beneath round-headed 5-light Perpendicular window with human and grotesque head stops. Twin round-headed belfry windows to each face beneath battlemented parapet with 8 crocketed pinnacles. Inscription on north face reads "DIEV TEMPLE YARD ET GARDE DE ROY".

GRADE: I

HERITAGE LIST NUMBER: 1295776

404 Driffield All Saints

LOCATION: Middle Street North, Great Driffield

POSTCODE: YO25 6SU

OS GRID REF: TA022579

DATE: C15

DESCRIPTION: West tower of 3 stages, approximately 34 metres high. Moulded plinth, string-courses to first and belfry stages. Angle buttresses with ogee-headed canopied niches, panels with blank cusped tracery, and crocketed gablets at upper levels. First floor, south elevation: pointed window with Y-tracery, under hood-mould with carved stops. Belfry openings of 3 transomed lights with Perpendicular tracery under hood-moulds with stops and crocketed ogee gablets. Battlemented parapet with traceried panels and 8 crocketed pinnacles. West elevation: west door with continuous shallow roll-moulding under crocketed ogee label with foliated finial flanked by heraldic shields. Pointed 4-light west window under crocketed ogee hood-mould with foliated finial flanked by small heraldic shields. Niche with moulded base and crocketed canopy to first floor, under carved panel with angels carrying heraldic shield.

GRADE: I

HERITAGE LIST NUMBER: 1083378

405 Easington All Saints

LOCATION: The Square, Easington

POSTCODE: HU12 0TS

OS GRID REF: TA398191

DATE: C14

DESCRIPTION: 3-stage tower: moulded plinth, full-height diagonal buttresses with offsets, moulded string-courses between stages. First stage: pointed 3-light west window with Perpendicular tracery, moulded sill string-course and hood-mould returned as string-course. Second stage: small round-headed 1-light trefoiled windows to south and west, line of earlier nave roof to east side. Pointed 2-light traceried belfry openings with hood-moulds. Moulded string-course, coped embattled parapet.

GRADE: I

HERITAGE LIST NUMBER: 1346602

406 Easingwold St John the Baptist & All Saints

LOCATION: Church Hill, Easingwold

POSTCODE: YO61 3JU

OS GRID REF: SE525700

DATE: C15

DESCRIPTION: 3-stage embattled tower which originally had a high open west porch of which the chamfered pointed arch remains, now blocked with a later door and window inserted.

GRADE: II*

HERITAGE LIST NUMBER: 1294335

407 East Ayton St John the Baptist

LOCATION: Main Street, East Ayton

POSTCODE: YO13 9HL

OS GRID REF: SE991849

DATE: C12

DESCRIPTION: 3-stage embattled tower. C12 with rebuilt upper stages. Single lancet to lowest stage to west, and 4 lancet bell-openings to upper stage.

GRADE: II*

HERITAGE LIST NUMBER: 1167755

408 Eastrington St Michael

LOCATION: High Street, Eastrington

POSTCODE: DN14 7PW

OS GRID REF: SE796299

DATE: C15

DESCRIPTION: 4-stage Perpendicular-style tower with stepped diagonal buttresses flanked by lean-to aisles. Chamfered plinth. Pointed west doorway of 3 moulded orders with hood-mould. 3-light Perpendicular window with hood-mould and head stops and 3 illegible coats of arms above. 2-light window with 4-centred arch to third stage of south facade, 2-light pointed Perpendicular belfry openings to fourth stage of each facade. Gargoyles, embattled parapet and 4 crocketed pinnacles.

GRADE: I

HERITAGE LIST NUMBER: 1031835

409 Ebberston St Mary

LOCATION: A170, 0.5 mile west of Ebberston village

POSTCODE: YO13 9PA

OS GRID REF: SE891833

DATE: Probably C14

DESCRIPTION: The 2-stage, embattled tower has 2-light louvred bell-openings with trefoil heads beneath pointed hood-moulds to all faces except the north, which has a single round-headed light. Similar opening to lower stage of west face.

GRADE: II*

HERITAGE LIST NUMBER: 1315780

410 Elloughton St Mary

LOCATION: Church Lane, Elloughton

POSTCODE: HU15 1SP

OS GRID REF: SE944282

DATE: C15

DESCRIPTION: West tower, plinth, diagonal buttresses with offsets. Slit windows to first stage. Chamfered first-floor band; 2-light pointed belfry openings with Y-tracery; eaves string-course; crenellated parapet with crocketed corner finials. West window: pointed, of 2 lights, with Perpendicular tracery.

GRADE: II*

HERITAGE LIST NUMBER: 1203258

411 Etton
St Mary

LOCATION: Main Street, Etton

POSTCODE: HU17 7PG

OS GRID REF: SE981435

DATE: C12

DESCRIPTION: C12 west tower (later raised) with clasping pilaster buttresses. Single round-headed slit window to lower stage; 2 round-headed belfry openings beneath corbel-table. Corbelled-out parapet above. Round-headed west door of 3 orders on nook-shafts with scallop capitals. Inner order: medallions with floral ornament; central order: beak-head; outer order: stylised beak-head. All under a filleted hood-mould with mask stops in a slight projection under a corbelled-out roof.

GRADE: II*

HERITAGE LIST NUMBER: 1103443

412 Folkton
St John the Evangelist

LOCATION: Village Street, Folkton

POSTCODE: YO11 3UH

OS GRID REF: TA053796

DATE: C13 with C19 additions

DESCRIPTION: 3-stage C13 tower with upper stage of 1877–81. Angle buttresses and staircase turret with slit windows to 2 lower stages; 2-light pointed bell-openings under pointed arches to third stage.

GRADE: II*

HERITAGE LIST NUMBER: 1148153

413 Fridaythorpe St Mary

LOCATION: Church Lane, Fridaythorpe

POSTCODE: YO25 9TA

OS GRID REF: SE874592

DATE: c.1190

DESCRIPTION: Tower with chamfered plinth. To west side, first stage, a round-arched window with chamfered jambs; above to second stage a wooden clock inscribed "ANO DOM / 1903 / Time is short : Eternity / is long"; to other sides, second stage, are rectangular bell-openings. Eaves band with C20 rainwater heads to east. Low parapet with pyramidal finials to corners.

GRADE: I

HERITAGE LIST NUMBER: 1346479

414 Garton in Holderness St Michael

LOCATION: Garton in Holderness

POSTCODE: HU11 4QB

OS GRID REF: TA270354

DATE: Early C13

DESCRIPTION: 2-stage west tower, double-chamfered plinth, lancets to first stage. Single lancet to belfry, west face.

GRADE: I

HERITAGE LIST NUMBER: 1215863

415 Garton on the Wolds St Michael

LOCATION: Station Road, Garton on the Wolds

POSTCODE: YO25 3ES

OS GRID REF: SE981593

DATE: C12 with C15 upper stage

DESCRIPTION: Tower has stepped and chamfered plinth and clasping full-height pilaster buttresses with offsets. To west a round-arched entrance: studded plank door in surround of 4 orders, nook-shafts have scallop capitals, the outer 3 arches have chevron moulding, the inner arch is roll-moulded, all under decorative hood-mould, and with patterned abaci and impost bands. Mass clock to right. Above are 3 eroded figures. Above again a decorative sill band and round-arched window between single nook-shafts with volute capitals, 2 arches of chevron moulding under continuous hood-mould. Above a slit window and then, as all other sides, 2-light 4-centred arched belfry opening with Perpendicular tracery to head and hood-mould. South side of tower has a round-arched window to first and second stages. North side has a round-arched window to second stage. Embattled parapets with stepped corner and central pinnacles.

GRADE: I

HERITAGE LIST NUMBER: 1160977

416 Gate Helmsley St Mary

LOCATION: Main Street, Gate Helmsley

POSTCODE: YO41 1NE

OS GRID REF: SE690552

DATE: C15

DESCRIPTION: 3-stage tower with blocked window, 3-light window with Perpendicular tracery to second stage, two 2-light belfry windows above surmounted by battlement with later pinnacles. Substantially restored 1885–86.

GRADE: II*

HERITAGE LIST NUMBER: 1149613

417 Goodmanham All Hallows

LOCATION: Main Street, Goodmanham

POSTCODE: YO43 3HG

OS GRID REF: SE889431

DATE: Late C12 with later additions

DESCRIPTION: West tower of 2 stages with chamfered string and later buttresses with offsets. West door, now blocked, with nook-shafts, scallop capitals, quirk and chamfer imposts ornamented with saltire crosses, and round arch with chevrons. 2-light square-headed belfry openings with 4-centred heads under hood-mould. Crenellated parapet with grotesques and gargoyles.

GRADE: I

HERITAGE LIST NUMBER: 1084132

418 Guisborough St Nicholas

LOCATION: Church Street, Guisborough

POSTCODE: TS14 6HG

OS GRID REF: NZ616161

DATE: c.1500

DESCRIPTION: Embattled tower of 3 stages has west door and window (glass by A. Fitzgerald Watt, 1908) under single giant arch.

GRADE: II*

HERITAGE LIST NUMBER: 1329545

419 Halsham All Saints

LOCATION: North Road, Halsham

POSTCODE: HU12 0DD

OS GRID REF: TA268277

DATE: C14

DESCRIPTION: 3-stage tower: angle buttresses with offsets to first stage, diagonal buttresses above; stair-lighting slits to south-west angle. First stage: pointed 3-light west window with Perpendicular tracery and hood-mould, small ogee-headed niche above with carved angel corbel and flanking pinnacled buttress shafts; blocked pointed arches, moulded string-courses to north and south. Second stage: small square-headed south window, moulded string-course. Pointed 2-light traceried belfry openings with wooden louvres. Moulded string-course, gargoyle to west, coped embattled parapet.

GRADE: I

HERITAGE LIST NUMBER: 1160776

420 Harpham
St John of Beverley

LOCATION: Crossgates, Harpham

POSTCODE: YO25 4QT

OS GRID REF: TA092615

DATE: Late C14

DESCRIPTION: 3-stage west tower: triple-chamfered plinth, angle buttresses with offsets, string-courses. Small square-headed window to second stage, pointed belfry openings with Y-tracery. Pointed double-chamfered west window with Perpendicular tracery.

GRADE: I

HERITAGE LIST NUMBER: 1083345

421 Hayton
St Martin

LOCATION: Back Lane, Hayton

POSTCODE: YO42 1RR

OS GRID REF: SE820460

DATE: C14, heightened later

DESCRIPTION: West tower of 3 stages with moulded plinth and string-courses. Second stage (originally the belfry) has pointed 2-light windows with curvilinear tracery. Upper stage has lancets under pierced crenellated parapet with corner pinnacles. 2-light pointed west window with C19 tracery.

GRADE: I

HERITAGE LIST NUMBER: 1084144

422 Healaugh
St John the Baptist

LOCATION: Wighill Lane, Healaugh

POSTCODE: LS24 8DA

OS GRID REF: SE498480

DATE: C12 with C19 additions

DESCRIPTION: 2-stage bell-tower, the second stage C19 with 2-light bell-openings. Corbel-table. Hipped roof.

GRADE: II*

HERITAGE LIST NUMBER: 1316655

423 Holme-on-Spalding-Moor
All Saints

LOCATION: A614, just north-east of Holme

POSTCODE: YO43 4EE

OS GRID REF: SE820389

DATE: C13 with C15 upper section

DESCRIPTION: 3-stage tower with lower stage of rubble, upper stages of ashlar. Diagonal stepped buttresses. Pointed doorway of 2 moulded orders with angel-head label stops, with 3-light transomed Perpendicular window with beast-head label stops above. String-course. Niche with nodding canopy containing seated kingly figure. String-course and 2-light transomed pointed Perpendicular belfry opening to each face. Pinnacles to diagonal buttresses. Gargoyles beneath openwork battlements with 8 pinnacles.

GRADE: I

HERITAGE LIST NUMBER: 1025868

424 Hornsea St Nicholas

LOCATION: 1 Newbegin, Hornsea town centre

POSTCODE: HU18 1AB

OS GRID REF: TA200476

DATE: Late C13 with later additions

DESCRIPTION: Late C13 west tower, altered and enlarged late C14 and C15. 3 stages, embraced by westward extension of nave aisles. Chamfered plinth, angle buttresses with offsets, chamfered strings. Pointed openings with Y-tracery under hood-moulds to second and third stages. Crenellated parapet with 8 crocketed pinnacles. Pointed double-chamfered west door under stopped hood-mould. 4-light west window with Perpendicular tracery under 4-centred arch.

GRADE: I

HERITAGE LIST NUMBER: 1249386

425 Hotham St Oswald

LOCATION: Main Street, Hotham

POSTCODE: YO43 4UF

OS GRID REF: SE894345

DATE: Early C12

DESCRIPTION: Broad, squat tower with round-arched west door, 2 roll-moulded orders on scalloped capitals within the soffit of the arch. 2-light cinque-cusped pointed window above breaks zigzag band which is continuous around all 3 exposed faces. Tower refaced above window. Round-arched belfry windows to each face. Tower restored in 1789.

GRADE: II*

HERITAGE LIST NUMBER: 1083305

426 Hovingham
All Saints

LOCATION: Church Street, Hovingham

POSTCODE: YO62 4LG

OS GRID REF: SE666757

DATE: C11

DESCRIPTION: 3-stage west tower articulated by square-section string-courses. Round-arched west door with free-standing angle shafts and 4 orders, the inner one recessed, the third with roll-moulding. C9 Anglian cross carved in high relief set into wall above doorway and below a single course of herringbone masonry. Second stage: double-splayed round-headed window to south; small rectangular slit windows immediately below string-course to west and north. Third stage: tall, narrow, double belfry windows to each face. C10 wheel-cross set into wall above south belfry window. Rebuilt corbel-table and roof.

GRADE: II*

HERITAGE LIST NUMBER: 1148990

427 Humbleton
St Peter

LOCATION: New Lane, Humbleton

POSTCODE: HU11 4NJ

OS GRID REF: TA226348

DATE: C14

DESCRIPTION: 3-stage west tower: diagonal buttresses, moulded strings. Slit windows to second stage, cusped Y-tracery to pointed belfry openings. Crenellated parapet. Pointed 3-light west window with Perpendicular tracery under hood-mould.

GRADE: I

HERITAGE LIST NUMBER: 1216050

428 Hunmanby All Saints

LOCATION: Church Hill, Hunmanby

POSTCODE: YO14 0JU

OS GRID REF: TA095774

DATE: C12 with C15 upper stage

DESCRIPTION: The unbuttressed square tower has slit windows to lowest stage, 2-light window under a rounded hood-mould to second stage, and 2-light bell-openings under square hood-mould to third stage.

GRADE: II*

HERITAGE LIST NUMBER: 1316442

429 Husthwaite St Nicholas

LOCATION: Main Street, Husthwaite

POSTCODE: YO61 4QA

OS GRID REF: SE518751

DATE: C15

DESCRIPTION: C12 stone reused in lower stages of tower which has plinth with chamfered offset, diagonal buttresses with many offsets to west; C15 ground-floor west window of 3 cinquefoiled lights in chamfered 4-centred arch; first-floor 1-light louvred south opening; string-course below C15 belfry with on each side a 2-light cinquefoiled louvred flat-headed chamfered opening; weathervane on roof.

GRADE: II*

HERITAGE LIST NUMBER: 1150764

430 Hutton Buscel
St Matthew

LOCATION: Fothill Lane, Hutton Buscel

POSTCODE: YO13 9LT

OS GRID REF: SE972840

DATE: C12 with C15 parapet

DESCRIPTION: 3-stage tower on plinth has a narrow round-headed lights to lowest stage on south face, and to second stage on west face. To each face, paired round-headed bell-openings with shafts beneath pointed hood-moulds; tympanum to each opening except that to south is pierced by a quatrefoil. Chamfered string-course to third stage. Corbel-table beneath embattled parapet pierced by stone waterspout.

GRADE: I

HERITAGE LIST NUMBER: 1172851

431 Hutton Cranswick
St Peter

LOCATION: Church Lane, Hutton Cranswick

POSTCODE: YO25 9PS

OS GRID REF: TA024533

DATE: C15

DESCRIPTION: Tower with stepped and chamfered plinth. South-west stair-turret. Angle buttresses with offsets to full height to west. West end has 3-light pointed window with Perpendicular tracery to head under hood-mould, partly recut. First-stage band. Second stage has slits and 2-light, straight-headed Perpendicular belfry opening to each side. Hollow-moulded band. Battlements with crocketed pinnacles to angles.

GRADE: II*

HERITAGE LIST NUMBER: 1161006

432 Keyingham St Nicholas

LOCATION: Church Lane, Keyingham

POSTCODE: HU12 9SX

OS GRID REF: TA245255

DATE: C13, partly rebuilt late C14–early C15

DESCRIPTION: 4-stage tower: plinth, angle buttresses with offsets to first 3 stages, moulded string-courses between stages. First stage: small blocked pointed hollow-chamfered north door. Second stage: pointed west window with restored Y-tracery, original hood-mould and head stops. Plain third stage with line of former nave gable to east. Pointed belfry openings with remains of former Y-tracery, hood-moulds and head stops, C20 wooden louvres. Coped parapet.

GRADE: I

HERITAGE LIST NUMBER: 1160841

433 Kilham All Saints

LOCATION: Church Road, Kilham

POSTCODE: YO25 4RQ

OS GRID REF: TA064644

DATE: C15

DESCRIPTION: West tower of 3 stages with hollow-chamfered plinth and diagonal buttresses. Pointed belfry openings of 2 lights with cusped Y-tracery under stopped hood-moulds. Crenellated parapet. Pointed west door with hood-mould; pointed 3-light west window with Perpendicular tracery under stopped hood-mould. Heraldic shield over, beneath crocketed ogee canopy.

GRADE: I

HERITAGE LIST NUMBER: 1309840

434 Kirby Underdale
All Saints

LOCATION: Main Street, Kirby Underdale

POSTCODE: YO41 1QY

OS GRID REF: SE808585

DATE: C11

DESCRIPTION: Mid-C12 west door of 2 round orders on nook-shafts with scallop capitals: chevrons to outer order, broad roll-moulding to inner. Immediately above the door is a small oblong window. Belfry openings of 2 square-headed mullioned lights. Crenellated parapet.

GRADE: I

HERITAGE LIST NUMBER: 1083837

435 Kirk Ella
St Andrew

LOCATION: Packman Lane/Church Lane, Kirk Ella

POSTCODE: HU10 7TL

OS GRID REF: TA020297

DATE: 1450–54

DESCRIPTION: West tower with high plinth, diagonal buttresses. West elevation has pointed door with continuous hollow chamfer beneath a 3-light pointed window with Perpendicular tracery under hood-mould. Canopied niche containing a statue of Saint Andrew holding his cross. Angel with shield supporting a mid-wall pilaster buttress flanked by gargoyles over the belfry opening. 3-light 4-centred arched belfry openings with Perpendicular tracery, crenellated parapet with 8 crocketed finials.

GRADE: I

HERITAGE LIST NUMBER: 1103401

436 Kirkburn St Mary

LOCATION: Kirkburn

POSTCODE: YO25 9DU

OS GRID REF: SE979551

DATE: C12–C15

DESCRIPTION: Tower with stepped and chamfered plinth. Angle buttresses with offsets up to second stage. To west a round-arched light with chamfered surround, slit window. First-stage band. To second stage, west side a slit window. Second-stage band. To third stage (C13) each side are 2-light Early English belfry openings. To fourth stage (C15) each side are 2-light, pointed Perpendicular belfry openings. Corbel-table. Low parapet.

GRADE: I

HERITAGE LIST NUMBER: 1083797

437 Langtoft St Peter

LOCATION: Church Lane, Langtoft

POSTCODE: YO25 3TN

OS GRID REF: TA008670

DATE: C13

DESCRIPTION: West tower of 3 stages: double-chamfered plinth, clasping and pilaster buttresses, roll-moulded string to belfry stage. Lancet to ground floor. Twin pointed belfry openings with mid-wall shaft, chamfered imposts, and hood-moulds. Former corbel-table below later parapet with moulded coping.

GRADE: I

HERITAGE LIST NUMBER: 1083356

438 Lastingham
St Mary

LOCATION: Front Street, Lastingham

POSTCODE: YO62 6TN

OS GRID REF: SE728904

DATE: C14

DESCRIPTION: 2-stage embattled tower on double-chamfered plinth with diagonal offset buttresses. To lower stage, west window of 2 trefoil-headed lights beneath panel tracery in square-headed surround. Bell stage openings are paired louvred lights with trefoil heads beneath flat hood-moulds. Moulded string-courses between stages and beneath parapet.

GRADE: I

HERITAGE LIST NUMBER: 1316041

439 Leake
St Mary

LOCATION: A19 Leake

POSTCODE: YO7 4BN

OS GRID REF: SE433906

DATE: C12

DESCRIPTION: Tower of 3 stages, flanked by aisles. Later pointed-arch west window of 3 cusped lights with hood-mould under band; to each aisle a 2-light Y-tracery pointed-arch window. To second stage on south side a 1-light arched window. Third stage has triple arcade to each side with cushion capitals to columns, blind outer arches and louvred 2-light belfry opening to central arches, corbel-table. Plain chamfered parapet.

GRADE: I

HERITAGE LIST NUMBER: 1315093

440 Lockton
St Giles

LOCATION: Church View, Lockton

POSTCODE: YO18 7PX

OS GRID REF: SE844900

DATE: C15

DESCRIPTION: 2-stage, embattled tower with diagonal buttresses has 2-light louvred bell-openings with foiled square heads to all faces except the north. West window of 2 trefoil-headed lights within a pointed arch. The tower has a saddleback roof.

GRADE: II

HERITAGE LIST NUMBER: 1260741

441 Long Marston
All Saints

LOCATION: Angram Road, Long Marston

POSTCODE: YO26 7NB

OS GRID REF: SE505508

DATE: C15

DESCRIPTION: Tower has 3-light west window, clock above; 2-light, 4-centred arched openings to second and belfry stages. String-course, battlemented parapet with gargoyles and crocketed finials. Staircase tower attached to south side.

GRADE: I

HERITAGE LIST NUMBER: 1150327

442 Lund
All Saints

LOCATION: The Green, Lund

POSTCODE: YO25 9TF

OS GRID REF: SE970481

DATE: C15

DESCRIPTION: Tower has a moulded plinth, diagonal buttresses with offsets. Small trefoil-headed light to lower stage. 2-light pointed belfry opening with cinquefoil heads under hood-mould with face stops and an angel at the apex. Crenellated parapet. 3-light pointed west window with Perpendicular tracery under hood-mould with grotesque stops. Above, the figure of a praying priest.

GRADE: II

HERITAGE LIST NUMBER: 1103410

443 Malton
St Michael

LOCATION: Market Place, Malton

POSTCODE: YO17 7LP

OS GRID REF: SE785717

DATE: C15

DESCRIPTION: 3-stage tower with angle buttresses, on chamfered plinth. C19 roll-moulded west doorway in 4-centred arch of 3 orders, beneath flat hood-mould with mouchettes in the spandrels. Inserted Perpendicular window over doorway. Belfry has pointed cusped openings beneath flat hoods, paired on north, south and west faces, single to east. Moulded strings between first and second stages on north and south faces interrupted on west by later window: moulded strings between second and belfry stages, and below plain parapet, on all sides.

GRADE: II*

HERITAGE LIST NUMBER: 1201930

444 Middleton St Andrew

LOCATION: Main Street, Middleton

POSTCODE: YO18 8NS

OS GRID REF: SE782854

DATE: C11 with C13 bell stage

DESCRIPTION: Tower of coursed rubble sandstone with long and short quoins. Blocked round-arched west doorway within raised surround, both with imposts. Crown of outer arch broken by an inserted vesica-shaped diamond-latticed window. 2 similarly glazed stair lights to south face of tower. Bell stage set back over string-course, with angle lesenes to south and west. Louvred bell-openings of paired pointed lights with centre shaft, recessed beneath chamfered pointed arches with nook-shafts. On north face the bell-opening is a lancet beneath ogee hood-mould which continues around the tower. Embattled parapet over corbel-table with waterspouts to north and south. Fragments of Anglo-Danish sculpted stone incorporated in the tower include a cross head and cross shaft, and a small animal.

GRADE: I

HERITAGE LIST NUMBER: 1315705

445 Monk Fryston St Wilfrid of Ripon

LOCATION: Main Street, Monk Fryston

POSTCODE: LS25 5DX

OS GRID REF: SE505297

DATE: C11 with C14 and C15 additions

DESCRIPTION: 3-stage tower: early C14 diagonal buttresses with offsets to first stage. 2-light west window with reticulated tracery. First stage band. 2-light round-headed bell-openings with shaft between to second stage. 2 corbel-tables. Third stage: 2-light pointed bell-openings with reticulated tracery to each side. Battlements with pinnacles.

GRADE: I

HERITAGE LIST NUMBER: 1296769

446 Nafferton All Saints

LOCATION: Westgate, Nafferton

POSTCODE: YO25 4JA

OS GRID REF: TA055590

DATE: C15

DESCRIPTION: Tower has moulded plinth, diagonal buttresses with offsets. 2-light pointed belfry openings with cusped tracery under stopped hood-moulds. Embattled parapet, crocketed corner pinnacles. West door under 4-centred head with continuous mouldings and stopped hood-mould. West window: pointed, of 3 lights, with Perpendicular tracery under hood-mould with beast stops.

GRADE: I

HERITAGE LIST NUMBER: 1346661

447 Newton Kyme
St Andrew

LOCATION: End of Croft Lane, Newton Kyme

POSTCODE: LS24 9LR

OS GRID REF: SE465448

DATE: C15

DESCRIPTION: 2-stage tower: plinth. 2-light, straight, trefoil-headed window. First-floor band. Slit window to south. Trefoil-headed, 2-light bell-openings. Band with gargoyles. Battlements.

GRADE: I

HERITAGE LIST NUMBER: 1132464

448 North Cave
All Saints

LOCATION: Church Street, North Cave

POSTCODE: HU15 2LW

OS GRID REF: SE896327

DATE: Late C12 and C15

DESCRIPTION: 2-stage tower: lower part late C12; upper part C15. 2-light Y-traceried west window with lancet above and diagonal buttress to left, and additional lean-to boiler house to extreme left. Clock face to north, west and south faces of tower with coved band above. 3-light Perpendicular belfry openings to each face. Coved band with gargoyles, embattled parapet with 8 crocketed pinnacles.

GRADE: I

HERITAGE LIST NUMBER: 1203419

449 North Dalton All Saints

LOCATION: B1246, North Dalton

POSTCODE: YO25 9UX

OS GRID REF: SE934522

DATE: C15

DESCRIPTION: West tower of 3 stages with moulded plinth and string-courses, and diagonal buttresses. Pointed 2-light belfry openings with ogee tracery under hood-moulds with heraldic stops; cusped running scroll under pierced crenellated parapet. C19 corner pinnacles. 2-light pointed west window with Perpendicular tracery including 2 brattished transoms. Tower restored 1872.

GRADE: II*

HERITAGE LIST NUMBER: 1161951

450 North Grimston St Nicholas

LOCATION: Main Street, North Grimston

POSTCODE: YO17 8AX

OS GRID REF: SE841678

DATE: C13

DESCRIPTION: Tower with clasping buttresses. To west: lancet window with evidence for former triangular hood-mould. Early C13 sculpture of Saint Nicholas, with tunic of tightly folded drapes held over one arm, the other holding a staff. String-course. Pointed double belfry openings. String-course to plain coped parapet. Weathering line of earlier steeply pitched roof visible on east face of tower.

GRADE: I

HERITAGE LIST NUMBER: 1174500

451 Old Malton St Mary

LOCATION: Town Street, Old Malton

POSTCODE: YO17 7HB

OS GRID REF: SE798725

DATE: Late C12

DESCRIPTION: South-west tower of 3 stages, on double-chamfered plinth, with clasping pilaster buttresses, the south-west one with inset angle shaft rising full height of tower. 2 lower stages of tower west face have single lancet windows, both of 3 roll-moulded orders, on detached shafts with water-leaf capitals, and shaft rings to second-stage window. Inner order on ground stage is enriched with continuous ball-flower moulding, and on second stage is chamfered. Both windows have leaf-stopped hood-moulds. Belfry openings are paired lancets of 2 orders beneath dog-tooth-moulded hood on leaf stop. Above belfry openings are 2 roundels of dog-tooth moulding, pierced by quatrefoils. North-west buttress has blind lancet on each stage, similar to those on west front. Slit openings to vice in south-west buttress. On south face, ground stage has single round-headed light. Arcade of 3 lancets, outer ones blind, to second stage, beneath continuous hood-mould; similar arrangement to belfry, centre lancet blind, with dog-tooth hood-mould. Pierced roundels are repeated above belfry openings. To east, ground stage has blocked pointed arch to former south aisle, between multiple shaft responds, the south one bearing traces of leaf capital. Single round-headed light to second stage. Tripled lancets to belfry, the outer 2 blind, beneath continuous dog-tooth hood-mould, and pierced roundels. On the north face, second-stage round-headed light is blocked, and belfry opening is single lancet. Belfry openings have scalloped louvres. On the west face, dog-tooth-moulded string-course at second-stage level continues across west front, broken by inserted west window, and returns on south face. Plain string-course to belfry stage encircles tower. Tower surmounted by plain parapet over corbel-table, carved as unfurled leaf buds.

GRADE: I

HERITAGE LIST NUMBER: 1201925

452 Osmotherley St Peter

LOCATION: School Lane, Osmotherley

POSTCODE: DL6 3BW

OS GRID REF: SE455971

DATE: C15

DESCRIPTION: West tower of 3 stages. Plinth, diagonal offset buttresses. South side second stage has small ogee-arched 1-light window, moulded band. Belfry: 2-light flat-headed mullion and transom openings with arched cusped lights. Eaves band, embattled parapet, pinnacles to each corner. To west side, a large C16 3-light mullion and transom window with arched lights, very flat 4-centred arch to head. Below band a clock face.

GRADE: II*

HERITAGE LIST NUMBER: 1188638

453 Pocklington All Saints

LOCATION: The Pavement, Pocklington

POSTCODE: YO42 2AS

OS GRID REF: SE802489

DATE: C15

DESCRIPTION: 4-stage west tower with triple-chamfered plinth, angle buttresses, and moulded string-courses. 2-light window with Perpendicular tracery under segmental head with stepped hood-mould to third stage. 3-light pointed belfry opening, with Perpendicular tracery with transom, under stepped hood-mould. Crenellated parapet with 8 crocketed pinnacles. Double-leaf boarded west door in 4-centred opening with continuous mouldings and hood-mould; tall pointed 5-light window with Perpendicular tracery under hood-mould with grotesques to stops and apex.

GRADE: I

HERITAGE LIST NUMBER: 1162006

454 Preston All Saints

LOCATION: Main Street, Preston

POSTCODE: HU12 8SA

OS GRID REF: TA187306

DATE: C15

DESCRIPTION: 4-stage tower: moulded plinth, angle buttresses with offsets and crocketed ogee gablets, continued as diagonal buttresses to upper section of belfry; moulded string-courses between stages. First stage: pointed moulded door with crocketed ogee hood-mould with head stops and pair of blank shields in relief above. Second stage: trefoiled ogee niches to buttresses with brattished corbels and octapartite ribbed canopies with crocketed ogee hoods; pointed 4-light west window with Perpendicular tracery, moulded transom, and ogee hood-mould, continued as string-course, supporting a niche containing a worn seated figure holding a shield, the niche with carved paterae to the corbel, sexpartite ribbed canopy, moulded arch with carved spandrels, moulded string-course and tall crocketed pinnacle. Third stage: ogee-headed slit lights to north and south, clock faces to each side. Fourth stage: twin pointed 3-light Perpendicular-traceried belfry openings with continuous hood-moulds and central shaft rising between windows to moulded string-course. Gargoyles to angles (apart from south-west), coped parapet with chamfered round-headed openings, 8 crocketed pinnacles (that to south-west restored) with ogee gablets to the shafts.

GRADE: I

HERITAGE LIST NUMBER: 1083438

455 Rowley
St Peter

LOCATION: Next to Rowley Manor hotel

POSTCODE: HU20 3XR

OS GRID REF: SE976326

DATE: Early C13 and C15

DESCRIPTION: Early C13 west tower, raised in C15. Lancet to second stage, 2-light pointed belfry opening with Perpendicular tracery, crenellated parapet with corner finials. 2-light pointed west window with Perpendicular tracery.

GRADE: II

HERITAGE LIST NUMBER: 1347015

456 Rudby in Cleveland
All Saints

LOCATION: Rudby Bank, Hutton Rudby

POSTCODE: TS15 0EY

OS GRID REF: NZ471066

DATE: C15

DESCRIPTION: Tower of 3 stages with diagonal buttresses, battlemented parapet and corner pinnacles.

GRADE: I

HERITAGE LIST NUMBER: 1150240

457 Rudston
All Saints

LOCATION: School Lane, Rudston, Driffield

POSTCODE: YO25 4UY

OS GRID REF: TA097677

DATE: C12 with C19 belfry

DESCRIPTION: West tower of 3 stages. Small lancets to lower stage, sill band and 2-light belfry openings with mid-wall shafts and typanum with circular sound holes all in round-headed openings with chamfered imposts and plain reveals. Pyramidal roof behind battlemented parapet on corbel-table.

GRADE: I

HERITAGE LIST NUMBER: 1162387

458 Salton
St John of Beverley

LOCATION: The Cottages, Salton

POSTCODE: YO62 6RN

OS GRID REF: SE716799

DATE: Late C12

DESCRIPTION: 2-stage tower with clasping buttresses, the south-west one housing the stair. North and south lancets to the lower stage, and C19 3-light west window with panel tracery. Louvred bell-openings recessed beneath round arches on nook-shafts. Continuous string-course below bell-openings; corbel-table below crenellated parapet. Pyramidal cap surmounted by filigree cross.

GRADE: I

HERITAGE LIST NUMBER: 1315726

459 Sancton All Saints

LOCATION: King Street, Sancton

POSTCODE: YO43 4RB

OS GRID REF: SE899394

DATE: C15

DESCRIPTION: Octagonal west tower of 2 stages: diagonal buttresses with offsets and crocketed pinnacles. Lower stage has pointed 3-light west window with Perpendicular tracery under hood-mould with trefoil-headed lancet above. 2-light pointed belfry openings with Perpendicular tracery, including double transoms, under hood-moulds. Crenellated parapet with crocketed pinnacles. Short flying buttresses to diagonal buttresses with pinnacles.

GRADE: II*

HERITAGE LIST NUMBER: 1084105

460 Saxton All Saints

LOCATION: Main Street, Saxton

POSTCODE: LS24 9PY

OS GRID REF: SE476369

DATE: Early C15 with restorations and additions of 1876 and 1907

DESCRIPTION: 2-stage tower has chamfered plinth incorporating remains of grave slab with incised cross and diagonal offset buttresses. To south side a narrow, round-arched plank door in chamfered surround. To west end a stepped 3-light window with cinquefoil heads under hood-mould. To upper stage a 2-light belfry opening with cinquefoil heads under hood-moulds. Corbel-table supports battlements with pinnacles.

GRADE: I

HERITAGE LIST NUMBER: 1168016

461 Settrington
All Saints

LOCATION: Church Lane, Settrington

POSTCODE: YO17 8NP

OS GRID REF: SE839702

DATE: C14

DESCRIPTION: 3-stage tower on offset plinth with chamfered string-courses to each stage. Diagonal offset buttresses. Pointed west doorway of 3 orders beneath head-stopped hood-mould. 3-light Perpendicular window above, and 2-light pointed bell-openings to third stage, all with head-stopped hood-moulds. Battlements pierced by quatrefoils over a band of flowing tracery and corbel-table of fleurons and masks. Carved waterspouts to north and south. Tiny grotesque figure set in the masonry between the window and the north-west buttress.

GRADE: II*

HERITAGE LIST NUMBER: 1173883

462 Sherburn in Elmet
All Saints

LOCATION: Church Hill, Sherburn in Elmet

POSTCODE: LS25 6AX

OS GRID REF: SE488335

DATE: C16

DESCRIPTION: Tower has massive diagonal offset buttresses, 2-light pointed bell-openings to each side with Perpendicular tracery to heads, and battlements with crocketed pinnacles.

GRADE: I

HERITAGE LIST NUMBER: 1148444

463 Sheriff Hutton St Helen & the Holy Cross

LOCATION: Church End, Sheriff Hutton

POSTCODE: YO60 6SX

OS GRID REF: SE657662

DATE: C12 lower part of tower, C15 upper part of tower

DESCRIPTION: Small round-headed window to left of porch roof, and blocked central round-headed window. Upper stage has twin belfry openings on string-course to each face, and battlemented parapet with pinnacles and gargoyles.

GRADE: I

HERITAGE LIST NUMBER: 1149591

464 Shiptonthorpe All Saints

LOCATION: Town Street, Shiptonthorpe

POSTCODE: YO43 3PE

OS GRID REF: SE852431

DATE: C13, heightened in C16

DESCRIPTION: 3 stages with moulded string to first stage and offset to belfry. First stage: lancet to west wall. Second stage: blocked slit windows to north, south and west walls. Double-chamfered square-headed belfry opening originally with central mullion. Crenellated parapet with 8 crocketed pinnacles.

GRADE: I

HERITAGE LIST NUMBER: 1310431

465 Sigglesthorne St Lawrence

LOCATION: Main Street, Sigglesthorne

POSTCODE: HU11 5QA

OS GRID REF: TA154457

DATE: Early C13, restored 1616 and 1763

DESCRIPTION: Low chamfered plinth, angle buttresses, chamfered string-courses. Lancet under hood-mould to first stage, central plaque to second stage with inscription "GG:TE" and sundial, with bronze 1616 "TA" gnomon, of 1818. Belfry openings of 2 trefoil-headed lights with attached shafts and vesica over in enclosing pointed arch with attached shafts and roll-moulding under hood-mould. Inscription (illegible) of 1763 to west elevation. Crenellated parapet.

GRADE: II*

HERITAGE LIST NUMBER: 1249769

466 Skerne St Leonard

LOCATION: Main Street, Skerne

POSTCODE: YO25 9HR

OS GRID REF: TA046551

DATE: C15

DESCRIPTION: West tower of 2 stages with moulded plinth, diagonal buttresses, hollow-chamfered string-courses. 2-light pointed belfry openings with C19 tracery in Perpendicular style. Embattled parapet with corner obelisks. 3-light west window with C19 tracery in Perpendicular style under hood-mould.

GRADE: I

HERITAGE LIST NUMBER: 1084138

467 Skipsea All Saints

LOCATION: Beeford Road, Skipsea

POSTCODE: YO25 8TG

OS GRID REF: TA165549

DATE: C15

DESCRIPTION: Chamfered plinth. Diagonal buttresses with offsets. To west end a 3-light pointed window with Perpendicular tracery to head, with recut hollow-chamfered mullions and within chamfered surround. Slit window. Boiler house to north. Second stage has slit window to south. Band. 2-light bell-opening with Perpendicular tracery to heads within pointed double-chamfered surrounds. Low parapet.

GRADE: I

HERITAGE LIST NUMBER: 1083824

468 Skipwith St Helen

LOCATION: Main Street, Skipwith

POSTCODE: YO8 5SQ

OS GRID REF: SE657385

DATE: C9–C11, with C15 upper stage

DESCRIPTION: First 2 stages have slit windows. First- and second-stage bands. To third stage are 2-light, trefoil-headed bell-openings. Quoins and battlements with pinnacles.

GRADE: I

HERITAGE LIST NUMBER: 1148467

469 Skirlaugh St Augustine

LOCATION: Church Lane, Skirlaugh

POSTCODE: HU11 5EU

OS GRID REF: TA141397

DATE: c.1401

DESCRIPTION: Diagonal buttresses with offsets, and crocketed pinnacles to top of second stage. Chamfered plinth with quatrefoil band. To west end hollow-moulded sill band. 3-light pointed window in double-chamfered surround with hollow-chamfered hood-mould with label stops which continue as band to sides. Second stage: brattished band. To west side a niche for a statue. Band. 2-light, pointed bell-openings with blind trefoil-headed lights below transom level. Band with fleurons and gargoyles. Pierced, traceried and crocketed battlements with crocketed pinnacles.

GRADE: I

HERITAGE LIST NUMBER: 1161844

470 Stainton in Cleveland St Peter & St Paul

LOCATION: Hemlington Road, Stainton in Cleveland

POSTCODE: TS8 9AY

OS GRID REF: NZ481141

DATE: C15

DESCRIPTION: Tower has paired cusped, louvred bell-openings with stone transoms. Embattled parapet with squat ogee-domed finials at angles. 1935 3-light west window, with cusped heads and hood-mould with scrolled stops, above C18 round-headed doorway. Boarded and studded double doors with scrolled strap hinges, in chamfered and roll-moulded surround under hood-mould.

GRADE: II*

HERITAGE LIST NUMBER: 1137540

471 Stillingfleet St Helen

LOCATION: York Road, Stillingfleet

POSTCODE: YO19 6SJ

OS GRID REF: SE593410

DATE: C13 lower stage with C15 and C16 additions

DESCRIPTION: Short angle buttresses. Single lancet to west. First-stage band. To south side a round-arched recess contains small mullion window. 2-light, straight-headed bell-openings. Battlements with pinnacles.

GRADE: I

HERITAGE LIST NUMBER: 1296904

472 Stillington
St Nicholas

LOCATION: Main Street, Stillington

POSTCODE: YO61 1JS

OS GRID REF: SE583678

DATE: Late C15

DESCRIPTION: Battlemented tower with 2-centred 2-light bell stage. Much restored and rebuilt.

GRADE: II*

HERITAGE LIST NUMBER: 1281543

473 Stokesley
St Peter & St Paul

LOCATION: High Street, Stokesley

POSTCODE: TS9 5AD

OS GRID REF: NZ526086

DATE: Mid-C14

DESCRIPTION: Perpendicular-style tower of 3 stages, with diagonal buttresses and battlemented parapet.

GRADE: II

HERITAGE LIST NUMBER: 1315420

474 Stonegrave Holy Trinity

LOCATION: Main Street, Stonegrave

POSTCODE: YO62 4LJ

OS GRID REF: SE655778

DATE: Early C12

DESCRIPTION: 3-stage tower. C15 2-light window in place of blocked round-headed west doorway. Small round-headed window in second stage of south face, splayed through full thickness of wall, its head being cut from a single stone. Third stage is later addition with twin square-headed transomed belfry openings to each face. Embattled parapet, the pinnacles being a Victorian addition. The tower is built over and against the west wall of the nave, thus suggesting an earlier date for the nave.

GRADE: II*

HERITAGE LIST NUMBER: 1173360

475 Sutton-on-Hull St James

LOCATION: Church Street, Sutton-on-Hull

POSTCODE: HU7 4TL

OS GRID REF: TA118330

DATE: c.1400

DESCRIPTION: Embraced west tower of 2 stages with moulded plinth, angle buttresses, quoins, string-courses and crenellated parapet with corner pinnacles. Moulded deeply recessed west door with hood-mould, above it a 5-light lancet with panel tracery and hood-mould. At the head of the lower stage, a small loop. To south, a clock. Bell stage has on each side a double-lancet louvred bell-opening with chamfered reveal. Tower restored, parapet and pinnacles renewed in 1886.

GRADE: I

HERITAGE LIST NUMBER: 1293238

476 Sutton-on-the-Forest All Hallows

LOCATION: Main Street, Sutton-on-the-Forest

POSTCODE: YO61 1DR

OS GRID REF: SE583647

DATE: Late C15

DESCRIPTION: South-east view has small, chamfered windows; a narrow chamfered light to second storey and a 2-light belfry opening on each side; clock face to south side, 3-light window with hood-mould to west side, blocked pointed-arch opening to north side; corner gargoyles to embattled parapet which has crocketed finials added 1877.

GRADE: II

HERITAGE LIST NUMBER: 1241818

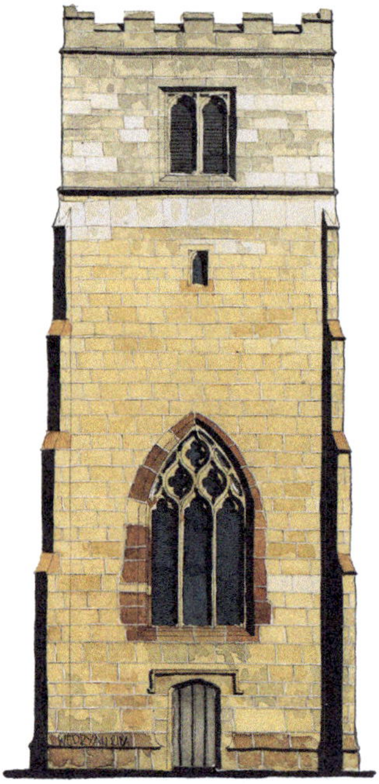

477 Sutton upon Derwent St Michael & All Angels

LOCATION: Main Street, Sutton upon Derwent

POSTCODE: YO41 4DQ

OS GRID REF: SE705473

DATE: C15

DESCRIPTION: West tower with plinth, diagonal buttresses. Trefoil-headed light under square head to second stage, 2-light trefoil-headed belfry openings. Crenellated parapet. External stair projection lit by small oblong windows, to south-east quoin. Pointed west window with reticulated tracery.

GRADE: I

HERITAGE LIST NUMBER: 1162085

478 Tadcaster St Mary

LOCATION: Kirkgate, Tadcaster

POSTCODE: LS24 9BL

OS GRID REF: SE485435

DATE: Early C14–early C15, taken down in 1875–77, re-erected and raised

DESCRIPTION: Diagonal buttresses with offsets surmounted by pinnacles. Round-arched west doorway in moulded surround under hood-moulds with head stops. Above a 5-light window with recut panel tracery under hood-mould with angel stops. South side has badly weathered canopied niche for statue which breaks first-floor band. Second-stage band. Twin 2-light bell-openings to each side with small buttresses with offsets rising between each pair supported by griffins and surmounted by battlements and pinnacles.

GRADE: II*

HERITAGE LIST NUMBER: 1167462

479 Terrington All Saints

LOCATION: Church Lane, Terrington

POSTCODE: YO60 6PS

OS GRID REF: SE671707

DATE: C15

DESCRIPTION: Tudor-arched doorway beneath triangular relieving arch, and 3-light C19 window with Perpendicular tracery to west front. Diagonal buttresses with offsets. Embattled parapet with pinnacles.

GRADE: I

HERITAGE LIST NUMBER: 1315764

480 Thirsk
St Mary

LOCATION: Kirkgate, Thirsk

POSTCODE: YO7 1PR

OS GRID REF: SE427823

DATE: C15

DESCRIPTION: West tower has large 3-light windows and bell-openings. 2-storey nave has 3-light depressed arched windows and panel tracery. The 3-light aisle windows have 2-centred arches. The single-storey chancel has 3-light basket-arched windows. 5-light east window.

GRADE: I

HERITAGE LIST NUMBER: 1314935

481 Thornton Dale
All Saints

LOCATION: Church Hill, Thornton Dale

POSTCODE: YO18 7QH

OS GRID REF: SE838831

DATE: Late C14

DESCRIPTION: 4 stages with battlements, crocketed pinnacles and diagonal buttresses with offsets. 2-light mullion and transom louvred bell-openings to each face beneath pointed hood-moulds which form an impost band. 3-light west window with reticulated tracery beneath pointed hood-mould on corbel heads.

GRADE: II*

HERITAGE LIST NUMBER: 1241098

482 Tunstall
All Saints

LOCATION: Main Street, Tunstall

POSTCODE: HU12 0JF

OS GRID REF: TA305320

DATE: C14, raised in C15

DESCRIPTION: West tower with moulded plinth, chamfered strings. Small lancet to second stage, 2-light pointed belfry openings with cusped Y-tracery. Crenellated parapet. 3-light pointed west window with Perpendicular tracery under hood-mould.

GRADE: I

HERITAGE LIST NUMBER: 1216255

483 Walkington
All Hallows

LOCATION: Little Weighton Road, Walkington

POSTCODE: HU17 8SL

OS GRID REF: SE998368

DATE: C15

DESCRIPTION: Chamfered and moulded plinth with moulded string-courses and diagonal buttresses with offsets. North side has pierced quatrefoil to first floor: above it, a canopied niche with crocketed finials. 2-light pointed belfry opening with Perpendicular tracery under a hood-mould. Crenellated parapet with crocketed corner finials. 3-light pointed west window with Perpendicular tracery. Clock face above quatrefoil to west wall.

GRADE: II*

HERITAGE LIST NUMBER: 1161425

484 Walton St Peter

LOCATION: Main Street, Walton

POSTCODE: LS23 7DJ

OS GRID REF: SE441478

DATE: C12 with C14 top stage

DESCRIPTION: Tower has chamfered plinth and offsets; no buttresses. West side has small lancet window beneath square-headed, chamfered window; offset below belfry stage having south clock and square-headed, 2-light openings in double-chamfered surrounds. Corbel-table and cornice below Perpendicular embattled parapet with 4 crocketed pinnacles.

GRADE: II*

HERITAGE LIST NUMBER: 1115540

485 Wawne St Peter

LOCATION: Main Street, Wawne

POSTCODE: HU7 5XW

OS GRID REF: TA091368

DATE: Early C13, raised in C15

DESCRIPTION: North-west tower with moulded plinth and angle buttresses with offsets. Ground floor has 2-light square-headed window with Perpendicular tracery, second stage has a clock, belfry stage has 2-light pointed openings with Perpendicular tracery and brattished transoms. West elevation: lancet to ground floor beneath carved human head, narrow trefoil-headed pointed window to second stage. Low parapet.

GRADE: I

HERITAGE LIST NUMBER: 1103427

486 Weaverthorpe St Andrew

LOCATION: Sherburn Road, Weaverthorpe

POSTCODE: YO17 8ES

OS GRID REF: SE966711

DATE: C12

DESCRIPTION: 4-stage tower and semicircular south-east vice, with slit openings. Twin round-arched bell-openings on central column with cushion capital, recessed beneath round arches with plain imposts. String-courses beneath belfry and plain parapet.

GRADE: I

HERITAGE LIST NUMBER: 1175527

487 Welwick St Mary

LOCATION: Main Street, Welwick

POSTCODE: HU12 0RX

OS GRID REF: TA341211

DATE: Late C13

DESCRIPTION: 2-stage tower with moulded plinth, quoins, diagonal buttresses with offsets to first stage. Square-headed 2-light west window with hood-mould and pointed relieving arch; pointed Y-traceried belfry openings, moulded string-course, coped parapet.

GRADE: I

HERITAGE LIST NUMBER: 1366256

488 Wetwang St Nicholas

LOCATION: Main Street, Wetwang

POSTCODE: YO25 9XJ

OS GRID REF: SE932590

DATE: C14

DESCRIPTION: Tower with stepped and chamfered plinth. Clasping buttresses with offsets. To first stage, west side, a lancet window. First-stage band. 2-light geometric traceried belfry openings to north, south and west sides within round-arched surrounds, to east side are 2 lancet openings. Corbel-table of masks and flowers. Projecting embattled parapets with band of mouchette decoration. Pierced quatrefoils to merlons. Stair-turret to south-east corner of tower has blocked slit windows.

GRADE: II*

HERITAGE LIST NUMBER: 1083774

489 Wharram-le-Street St Mary

LOCATION: Wharram-le-Street

POSTCODE: YO17 9TL

OS GRID REF: SE863659

DATE: C11 with C12 top stage

DESCRIPTION: 4-stage tower with blocked round-headed doorway of 2 moulded orders on narrow shafts with pendant triangle capitals, probably of Norman date. Square-headed slit windows to second and third stages. String-course. Double round-headed belfry windows. Rebuilt 4-course parapet, slightly corbelled out.

GRADE: I

HERITAGE LIST NUMBER: 1149064

490 Whenby St Martin

LOCATION: Whenby

POSTCODE: YO61 4SE

OS GRID REF: SE630698

DATE: C15

DESCRIPTION: Tower has hollow-chamfered plinth; stepped angle buttresses; blocked west doorway with continuously hollow-moulded pointed arch; ringing-floor window of 2 trefoiled lights with cusped quatrefoil above; 2-light flat-headed belfry opening with trefoil cusps; matching belfry opening to south; chamfered 1-light belfry opening to north.

GRADE: II*

HERITAGE LIST NUMBER: 1191384

491 Whitby St Mary

LOCATION: Abbey Plain, Whitby

POSTCODE: YO22 4JT

OS GRID REF: NZ901112

DATE: C12 and C13

DESCRIPTION: Tower of 3 stages; C15 embattled parapet.

GRADE: I

HERITAGE LIST NUMBER: 1055865

492 Whorlton
Holy Cross Old Church

LOCATION: Castle Bank, Whorlton

POSTCODE: DL6 3HT

OS GRID REF: NZ483025

DATE: C14

DESCRIPTION: Tower has pointed-arch inner door, ogee-headed bell-openings and plain blocking course below pyramidal roof with ball-on-spike finial.

GRADE: I

HERITAGE LIST NUMBER: 1294328

493 Wighill
All Saints

LOCATION: Church Lane, Wighill

POSTCODE: LS24 8BG

OS GRID REF: SE473465

DATE: C15

DESCRIPTION: Tower has plinth; cusped window in deeply chamfered surround to first stage, south side; each side has a belfry window of 2 cusped lights in shallow pointed arch; battlemented parapet with finely detailed gargoyles and crocketed pinnacles.

GRADE: I

HERITAGE LIST NUMBER: 1315379

494 Willerby
St Peter

LOCATION: Wain's Lane, Willerby

POSTCODE: YO12 4SH

OS GRID REF: TA008792

DATE: Late C14

DESCRIPTION: Squat tower with embattled parapet. Dwarf offset buttresses flank west window of 3 pointed lights with renewed chamfered mullions. Louvred 2-centred arched bell-openings in east, west and south faces. Pent-roofed vice on north face. String-course beneath parapet.

GRADE: II*

HERITAGE LIST NUMBER: 1175601

495 Wistow
All Saints

LOCATION: Church Hill, Wistow

POSTCODE: YO8 3UU

OS GRID REF: SE592356

DATE: C15

DESCRIPTION: Perpendicular 3-stage tower with stepped diagonal buttresses. Doorway to west has hood-mould with 2 demi-figures of angels as stops. Similar hood-moulds to window and bell-openings.

GRADE: I

HERITAGE LIST NUMBER: 1174776

496 York
All Saints, Pavement

LOCATION: High Ousegate, York

POSTCODE: YO1 8RZ

OS GRID REF: SE604517

DATE: c.1400 (lantern rebuilt 1837)

DESCRIPTION: 2-stage west tower surmounted by octagonal lantern. West window of 5 cinquefoiled lights beneath panel tracery in 2-centred head, flanked by offset buttresses. North and south faces have small pointed doorways to aisle roofs; clock face above each. Belfry openings on each face are of 3 cinquefoiled louvred lights in flattened 4-centred heads with hood-moulds, above belfry string. Lantern is buttressed with crocketed pinnacles and gargoyles between tiered lights of twin cinquefoiled openings with panel tracery in 2-centred head. Parapet composed of cinquefoiled openings beneath crocketed gablet.

GRADE: I

HERITAGE LIST NUMBER: 1256921

497 York
Holy Trinity

LOCATION: Goodramgate, York

POSTCODE: YO1 7LF

OS GRID REF: SE604520

DATE: Late C15

DESCRIPTION: Buttressed 3-stage tower on moulded plinth has 2-centred west window of 5 cinquefoiled lights with panel-traceried head in casement-moulded surround. Top stage has blocked opening to each face, of twin trefoiled lights with traceried 2-centred heads, the blocking pierced by glazed slits. String-course to top stage, and embattled parapet to rebuilt saddleback roof.

GRADE: I

HERITAGE LIST NUMBER: 1257686

498 York Holy Trinity

LOCATION: Micklegate, York

POSTCODE: YO1 6LE

OS GRID REF: SE598515

DATE: 1453

DESCRIPTION: Tower of 5 stages with embattled parapet and offset angle buttresses, north-east one on chamfered plinth. North-east and north-west buttresses are stop chamfered on 2 lowest stages. Ground stage incorporates part of original north aisle wall on chamfered plinth, and chamfered lancet window beneath 2-centred arch on jamb shafts with roll necking and nail-head moulded capitals. Hood-mould on floral stops and double-chamfered sill string. On north, east and west faces, belfry has a chamfered round-headed louvred opening recessed beneath round arch on jamb shafts with roll necking and moulded capitals. Moulded string at belfry level. West end buttressed with pilasters, north one original retaining vestigial twin gabled niches on west face, and one similar niche on north face with trefoiled head and nail-head moulded capitals.

GRADE: I

HERITAGE LIST NUMBER: 1257274

499 York
St Cuthbert

LOCATION: Peasholme Green, York

POSTCODE: YO1 7PW

OS GRID REF: SE607520

DATE: Mid-C15 incorporating late C11 east wall.

DESCRIPTION: 2-stage tower with south-west angle buttress: vestry projects to north. West window of 3 cinquefoiled lights with panel tracery in 4-centred head, beneath coved hood-mould; above, small trefoiled light. Belfry openings to each face are square-headed, of 2 louvred lights, beneath hood-moulds on weathered head stops. Moulded strings to belfry and beneath embattled parapet, with defaced gargoyles to both.

GRADE: I

HERITAGE LIST NUMBER: 1256888

500 York
St Lawrence

LOCATION: Hull Road, York

POSTCODE: YO10 3BN

OS GRID REF: SE612513

DATE: Late C12 with C13 alterations, top storey added C15

DESCRIPTION: Tower of the Old Church of St Lawrence. Roughly squared limestone rubble, the upper stage in larger and more regular blocks. Windows of bell stage infilled with brick. Doorway has a semicircular arch of 4 orders: the inner order has plain voussoirs springing from moulded imposts over scalloped capitals; the second order has an interlace above 2 monsters springing from capitals also carved with monsters; the 2 outer orders are carved with formalised foliage. Behind, a semicircular rear arch springing from chamfered imposts is probably the original tower arch of the late C12. On the south side the middle stage has 2 narrow blocked chamfered openings, the upper one with a pointed head. On the west side the lowest stage has a blocked Tudor-arched window, and the middle stage has a blocked pointed window. The north side has a blocked pointed window to the middle stage. Above and below the bell stage there are moulded string-courses. Gargoyles remain at the south-west and south-east corners, and the parapet is a C20 rebuilding with plain corner pinnacles. The bell-openings are of 2 trefoiled lights under straight heads with hood-moulds.

GRADE: I

HERITAGE LIST NUMBER: 1257511

Glossary

Anglo-Saxon/Saxon	pre-Norman style, C7–1066
ashlar	square-hewn stone
baluster	small pillar or pedestal of bellied form
batter	amount of variation from the perpendicular; a receding slope of a wall, structure or earthwork
belfry	bell space in church tower
billet	short roll inserted at intervals in hollow moulding
bolection	moulding raised above panel
boss	projection at intersecting point of vault
bracket	projecting piece of stone providing support to adjoining matter
brattishing	a decorative cresting at the top of a cornice or parapet
buttress	a projection from a wall, bonded to it to create support
castellated	battlemented; with crenellation
cavetto	moulding of quarter-round section
chamfer	surface formed by cutting off a square edge, usually at 45 degrees
chevron	V-shaped motif used in series to decorate moulding
clasping buttress	at corner of building on two elevations
colonnette	a small column or shaft

coping	the covering course, or capping, of a wall or parapet, to throw off rainwater
corbel	block of stone or timber projecting from wall to support weight
cornice	ornamental moulding around the top of a wall
course	horizontal layer of structural material
cove	concave moulding on large scale
crocket	leafy globe of stone, usually on sloping edges
cusp	projecting point in window-bar tracery
cyma moulding	ogee moulding of cornice
Decorated	historical division of English Gothic, mid-C13 to mid-C14
dog-tooth	moulding, a series of small pyramids formed by four leaves meeting to a point
dressing	smooth finish to stonework
dripstone	projecting stone moulding to discharge rainwater from face of building
Early English	historical division of English Gothic, mid-C12 to mid-C13
eaves	lowest edge of roof
embrasure	small opening in parapet of fortified building
faucet	spigot for controlling flow of rainwater from roof
finial	top feature, usually decorated
fleuron	decorative carved feature of leaves
foil	space between cusps of window
foliated	*see* fleuron
gable	area of wall at end of multi-pitched roof
gargoyle	waterspout projecting from parapet, often carved with human/animal shape
Gothic	period of English architecture using pointed arch, mid-C12 to mid-C15; *see also* Decorated; Early English; Perpendicular
herringbone	masonry or brickwork in zigzag courses
hipped roof	roof with pitched roof at gable end
hood-mould	*see* dripstone

jamb	one of the vertical sides of an opening
lancet	narrow pointed-arch window
lead	metal most commonly used for roofing, flashings and decorative work
lesene	a narrow, low-relief vertical pillar in a wall
lintel	horizontal beam taking load over opening
louvre	opening covered with horizontal sloped boards, spaced to allow ventilation
lozenge	diamond shape
machicolations	series of openings under a projecting parapet
merlon	solid portion of fortified parapet
mouchette	curved form of dagger-shaped window tracery
mullion	vertical pier that divides the lights of a window
nail head	ornamental motif moulding with small pyramids being regularly repeated
niche	vertical recess in wall, usually to receive sculpture
Norman	style of architecture preceding Gothic, 1066–c.1200
offset	sloping ledge in wall where thickness of part above is diminished
ogee	double curve, bending first one way and then another
order	one of a series of recessed arches and jambs forming a splayed opening
ovolo	wide convex moulding
parapet	a low wall, sometimes ornamented, at the edge of a wall, balcony or terrace
Perpendicular	historical division of English Gothic, mid-C14 to mid-C15
pier	stonework to carry a vertical load
pilaster	flat classical column in shallow relief against a wall
pillar	generally a free-standing supporting construction
pinnacle	tapering finial often decorated with crockets

plinth	projecting base course of a wall of a building
quoins	dressed stones at the corners of a building
rebate	usually rectangular recess formed in wall to receive door
reticulation	arranged in small squares
rubble	type of masonry, using irregular-shaped stone
saddleback roof	a normal pitched roof when used over a tower
saltire	cross with diagonal limbs
Saxon	*see* Anglo-Saxon
sill	piece of stone or timber forming the horizontal bottom of a door or window opening, designed usually to throw off rainwater
spandrel	triangular spaces between an arch and its enclosing rectangle
splay	chamfer, usually of a reveal
stops	plain or decorated block terminating a moulding
string-course	horizontal band of mouldings on a building
tracery	pattern of intersecting mullions and transoms in a window
transom	horizontal bars in windows, separating the lights
tympanum	space between a lintel and the arch above
vent	loophole in embattled wall
vesica	oval with pointed head and foot, usually of a window or tracery
vice	spiral staircase
voussoir	one of the wedge-shaped stones forming an arch

Index

THE EVOLUTION OF MEANING

1. A Universal Learning Process

Brendan Graham Dempsey

The Evolution of Meaning:
1. A Universal Learning Process
First Edition

Sky Meadow Press
Stannard, Vermont
www.skymeadowinstitute.org

ISBN 979-8-9903633-2-8

Contents

Other Books Published
by Sky Meadow Press:

Gurdjieff for a Time Between Worlds by Layman Pascal

Foreword

Sky Meadow Institute was founded in 2023 as an organization dedicated to *advancing systems-based thinking about the things that matter most.* That's a succinct, tidy way to try and express a whole bunch of interrelated concerns and efforts. So, what does it *really mean*? And how does a phrase like that try to tie together content as seemingly disparate as permaculture, psycho-social development, big history, metamodernism, and holistic well-being—topics core to the Institute's vision?

Systems are dynamic networks that unite many components into an integrated whole—one with new capacities and capabilities. Many elements working together can produce new powers and outcomes that transcend the mere sum of their parts. That's *synergy*; that's *emergence*—when increasingly complex, energy-unlocking configurations somehow allow for novel forms of depth, strength, vitality, and flourishing.

But if words like "system" and "complexity" sound too cold and mechanical, just replace them with something more organic. Think generative *relationships* and *relationality*, the *context-sensitive*, the *holistic.* That's what Nature is all about, after all. The *web of life*, *interconnection*, *symbiosis*, *Gaia.* The word "complexity" just means "to weave together," and that's what the tapestry of existence ultimately is: an ever more intricate embroidery revealing an ever more refined image of an ever-evolving cosmos.

Forging more intricate relationships is what the Universe does—and keeps doing, since out of those relationships come new entities with entirely new abilities which in turn create the possibility for even more novel relationships to cohere, and so on. So quarks relate to form atoms, atoms relate to form molecules, molecules relate to form cells, etc., all the way up to complex, conscious, caring human beings, who themselves relate to one another to form a vast global society rapidly coming to increasing knowledge of the cosmos.

Every "thing," then, is really a *process*: a dynamic relational event unfolding in relation to all other dynamic relational events. That's what the Universe is. That's what *you* are—a mind-blowing moment

at the edge of this 14-billion-year process of cosmic complexification and evolution, headed for still greater complexity…

Beautiful as this perspective is, it can be a hard one to adopt for most people. Complex systems are, well, *complex*, not simple, and that makes them difficult to see and make sense of. It's no accident that the field of complex systems science only got seriously underway in the last 50 years with the aid of supercomputers and advanced mathematics. We may intuit complexity, feel it in our bones and in our hearts, but to engage it conceptually with scientific rigor or apply its models to different domains of life with methodological clarity is no easy task.

Indeed, we shouldn't forget that simply getting comfortable with the basic sort of *linear*, *analytical* thought common to modern industrial cultures itself took centuries and required a vast reconfiguration of society just to take root. But while that sort of thinking has offered us incredible new insights into the nature of reality and unlocked the potential of expansive new science and technology, it's become painfully obvious that even *it* is not enough. Modern analysis simply misses too much of the Bigger Picture, with the result that we've become exceedingly skilled at a narrowly focused, objective-oriented form of thought that isn't taking into account all the relationships, interconnections, and complexity required for a proper understanding of impacts and outcomes. We are, in short, missing what matters most.

The results of this too-simplistic mindset are disturbingly more apparent by the day. The environment is rapidly degrading, biodiversity is tanking, human health is suffering, and social systems are crumbling. The old way of making sense of the world simply isn't cutting it anymore, and we'll need to "level up" our thinking yet again if we're to see ourselves into a flourishing 21st century.

That's what Sky Meadow Institute is dedicated to see happen. We need more systems-based thinking about the things that matter most—things central to our survival and well-being, like food, Nature, human relationships, governance, culture, and spirituality. Viewed through the lens of integration (not analysis), complexity (not commodity), systems (not dissection), our understanding and relationship to *all of these* shift in a radical and life-enhancing way.

When it does, our meaning-making itself complexifies. Our worldview becomes more expansive, sensitive, deep. We see more of that Bigger Picture, gain more avenues for connection and self-expression, and reduce the amount of suffering in the world that's caused by myopia and short-sightedness. Less and less can be truly "externalized" as the one relational Whole becomes all in all—you and me, us and them, everything, together.

Human civilization has never operated from such a perspective, yet likely cannot continue, in the end, unless some form of it begins to take hold and reshape how we live. Bold as such an aim may be, we can be heartened that, despite myriad backtracking and (not infrequent) disastrous relapses, the last 30,000 years of world history can be read as a sort of long-term psycho-social *learning process*. As social systems have complexified, bringing more and more of their subsystems together, they've unlocked new capacities and continually expanded people's worldviews in such a way as to allow us to appreciate more and more of the complexity of things.

Today, we need *a worldview of complexity* if we're to see civilization, if not life itself, to the end of the 21st century and beyond. Such a worldview sees in the complexification process itself a spiritually profound narrative. Moving beyond modern utilitarianism and postmodern nihilism, we must reframe issues of meaning and purpose within a broader cosmic frame of evolution and productive, care-enhancing integration.

The emerging metamodern paradigm seems to be speaking to such a shift in perspective. A truly metamodern spirituality is precisely that which brings a complexity lens to bear on the things that matter most. Part of that entails looking at the very evolution of worldviews and meaning structures as a process of complexification—a vital, promising, but (as yet) under-explored field of study. More broadly, it entails engaging complexification, integration, and emergence as fundamental dynamics of the self-organizing cosmos—processes in which we ourselves take part through our own journeys of learning, development, integration, and self-transcendence.

Seeing these deep patterns playing out across the fractal spans of self and Universe, stepping into a worldview of worldviews, and

learning about learning all help expand our perspective and return a profound sense of meaning to life after epochs of disenchantment and disillusionment. After the revelations of modern and postmodern thought, we are once more, and at long last, learning to find ourselves *at home in the universe* again—not through naïve projection and assimilation, but through genuine systemic integration into and as a singular, sublime Whole.

Though diverse, all of the books published by the Institute serve to inform and develop this line of thought in some way or another—whether that be by exploring "surplus cohesion" models of human spirituality and parts-work, advancing comprehensive unified theories of complexification and emergence, exploring new mythopoetic avenues for the development of the God-concept in an open-source and networked culture, etc. As a metamodern worldview of systems, complexity, and integrative pluralism comes online, Sky Meadow Press is dedicated to promoting and advancing the emerging discourse by publishing thinkers at the leading edge of this promising new paradigm. In time, we hope to see this effort blossom into a robust wisdom library for navigating this time between worlds and helping build bridges into the mist of an uncertain future.

The challenges before us are daunting, to be sure, and the path ahead will be hard if we're to see a successful update to the civilizational operating system. Still, despite the urgency and seriousness of the moment (or, indeed, because of them), we must move forward not just boldly but *joyfully*, marching ahead with a post-tragic perspective that sees the darkness but is not overcome by it. With levity, then, and a sincere irony emblematic of a new cultural sensibility, a new, weird and wild world is coming into view. I hope the books published here will help chart some of its emerging landscape and give us all a taste of what wonders lie ahead—if only we can rise to meet the crises of our time with courage and wisdom enough to greet them.

Brendan Graham Dempsey
Stannard, Vermont

A Universal Learning Process

ἀλλ᾽ ἐκδιδάσκει πάνθ᾽ ὁ γηράσκων χρόνος.
– *Prometheus Bound*

This book series is about the evolution of meaning. But what, exactly, is *meaning*, anyway? What does it really entail for something to be *meaningful*? Is it something purely subjective, relative—just a matter of taste or opinion? Or is anything ever objectively meaningful, *inherently* meaningful, necessarily so?

Clearly, meaning matters to people. Psychological research over the past few decades has revealed a direct association between people's sense of meaning in their lives and their overall health and well-being.[1] A perceived lack of it leads to pathology and, in extreme cases, suicide.[2] Meaning, then, has real existential import. But is this just a preoccupation of humans (perhaps only overly reflective, modern ones)? Or do other animals experience a sense of meaning? Is there meaning for a tree—or a rock? *What is the meaning of meaning?*

A comprehensive treatment of the evolution of meaning invites a consideration of etiologies and origins. And while this book series will focus primarily on human meaning and its transformations, I think it is vital to contextualize that focus by establishing the deep continuity that exists between human meaning and meaning more broadly. To that end, this book will consider meaning in its most fundamental terms, then follow this throughline up to the domain of meaning in culture.

Here I will attempt an excavation of meaning to its deepest strata. Such an archeology can allow us to trace the idea of meaning back, far before human society, and even before animal cognition or biological evolution. Meaning, I will argue, arises out of the most fundamental dynamics of the physical world: energy and information. Indeed, bold as it may sound, it is precisely via the thermodynamic forces driving cosmic evolution as a whole that meaning evolves into more complex biological and social registers such as we experience as conscious agents.

Meaning is a particular kind of information—namely, information linking entities with their environments in ways that causally impact their viability and flourishing. Meaning is "existential," then, indeed. Beginning with a consideration of energy and information at the most basic level, we will turn to developments in information theory that have helped clarify the nature of *meaningful information* specifically. I will then track the implications of this idea into the big history frameworks that map cosmic complexification, arguing that meaning itself evolves through new emergent planes of existence—from matter to life to mind and on into culture. As entities acquire meaningful knowledge through novel information processing systems at each of these levels, they engage in an ongoing learning process underway in the universe—a learning process, in fact, that characterizes *all* levels of cosmic evolution. After we get a sense of the full scope of this learning process, we can begin to zoom in on the cultural plane specifically to examine the evolution of human meaning, with particular attention to evolving conceptions of the sacred.

But first: the basics.

Energy and Information

There is a fundamental property of the universe that literally *in-forms* (i.e., gives form to) all known phenomena. This is the all-important Second Law of Thermodynamics. According to this law, nature tends toward equilibrium and the homogenization of differences in the cosmos. Heat dissipates, gradients dissolve, and "things fall apart." Distinct differences are inherently unstable and will face the natural push to even out—for, indeed, this "push" is precisely how energy does work.[3]

To picture this dynamic, consider what happens when you pour cream into your morning cup of coffee. It naturally diffuses, mixing with the coffee in the cup, so that the ordered distinction between "cream" and "coffee" gradually disappears, and both become inextricably mixed up. This is akin to the way energy in a system will dissipate and yield to *entropy*.

Figure 1.1. Entropy[4]

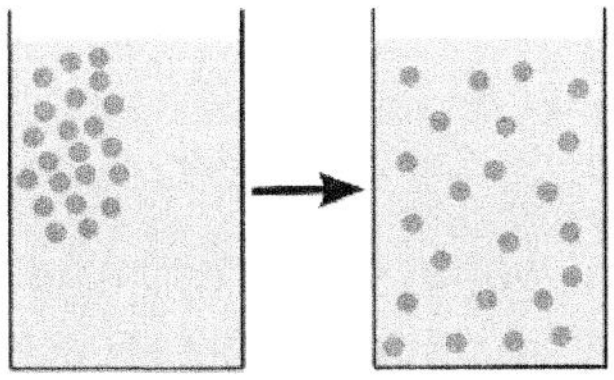

This natural tendency towards homogeneity was given a statistical account by the physicist Ludwig Boltzmann, who showed that the spontaneous trend towards equilibration could be understood in probabilistic terms. That is, in a system

with many possible microstates, highly ordered (e.g., concentrated) ones are simply far less likely than the vast range of ways they can be disordered. Entropy in this sense can be understood as the gravitation towards *randomness*, and equilibrium arises when chance and probability are the only structuring pressures on a system.

Now, if this entropic drift were the *only* dynamic operating in the universe, no structures or complex phenomena would ever form; there would only be static equilibrium—forever. Fortunately, this is not the case. Instead, ordered structures *can* arise, but *only if they are able to leverage the necessary energy to push back against the draw of entropy and maintain their improbable order.*

To make a common comparison, your house will tend to get messier and messier over time unless you make some *effort* to tidy up and organize it. There are a million places your dishes *can* be, so a clean kitchen means resisting that natural draw towards randomness and exerting yourself to ensure their configuration is in fact constrained to a very improbable range of overall possible states. Without that expenditure of energy, disorder will gradually increase and eventually you will find yourself living in "chaos." Things will mix and blend and blur, unless some energy is expended to keep entities distinct and give things form.

The same idea applies to information. Let's say your coffee mug has gone missing despite your best efforts to keep things organized. If I tell you, "It's not in the bathroom sink," you'll see that this fact is not all that informative. Why? Because there are so many places it *could* be that it not being in any one of them is actually highly probable. It would be different if the mug could only be in, say, one of two possible places—then

knowing it is not in the bathroom sink *is* incredibly informative. But with more possible states, there is more uncertainty, and thus more ignorance. If you told me, however, that my mug *is* in the bathroom sink, you will have conveyed a considerable amount of information indeed, since you will have narrowed down the vast range of possible facts to a single fact: the needle in the haystack of possible states, as it were.

In short, the more improbable the outcome, the more informative. This is basically the insight Claude Shannon offered that helped connect Boltzmann's notion of entropy to *information.*[5] If I roll two dice, one with 6 sides and one with 600, the outcome of the die with 600 will yield more bits of information, since the alternative number of outcomes was greater. A clean kitchen is quite informationally rich as there are many, many places my dishes could be but only a small, improbable subset in which they actually belong. Achieving that informational richness requires an expenditure of energy. (And, to be sure, a bigger kitchen or a more elaborate configuration of dishes would require even more energy.)

In sum, entropy jumbles distinct, informationally rich configurations, but energy can be deployed to *push back* against entropic randomization, maintaining order and information in the face of dissolution. When it does, differences are established and distinction is allowed to persist. The Second Law thus creates a fundamental necessity for any entity that would exist in the universe: it must utilize free energy to maintain itself in the face of entropy. That is the veritable "Prime Directive" for any thing to *be.* That is, for anything to exist, it must resist annihilation.

This seeming tautology is actually quite profound, as it provides a minimal explanation for the existence of everything in the universe. It is the basic logic that informs all the complex phenomena we see around us. Form requires energy to inform it—and the more structurally complex something is, the more energy it needs to resist the entropic draw towards formless equilibrium.

In the context of me and my kitchen, *I* am the means by which energy is being exerted to counter entropic disruption. The energy needed to keep the kitchen organized cannot come from the system of the kitchen itself (since energy in an isolated system entropically degrades according to the Second Law and loses the ability to do work); rather, the kitchen is an "open system," with energy flowing into it from the outside to do the work required to keep it organized and informationally rich.

Only open systems can resist entropic dissolution in this way. For instance, if you add hot water into the isolated system of a thermos of cold water, thermal equilibrium is inevitable. The energy will equilibrate and homogenize in accordance with the Second Law. Why? Because there is no additional energy available to do the work of keeping them distinct. If improbable configurations require energy to be sustained, then that energy will need to come from *outside* the system itself.

All things that exist, then, exist in relationship to the energy dynamics of their environment. All systems are systems-in-context. Indeed, without an energetic background context to distinguish and inform it, there could be no "system." As Santa Fe Institute's David Krakauer and colleagues frame this relationship through their *information theory of individuality*, systems are defined by their environments

through a figure–ground-style distinction.[6] Both represent information that bear on one another. Environments (E) influence systems (S), systems influence their environments, and each influences itself over time. Put in informational terms, there is a continual exchange of information from environment to system, from system to environment, from past to future states of the environment, and from past to future states of the system (Figure 1.2).

Figure 1.2. Flows of Information Between System and Environment[7]

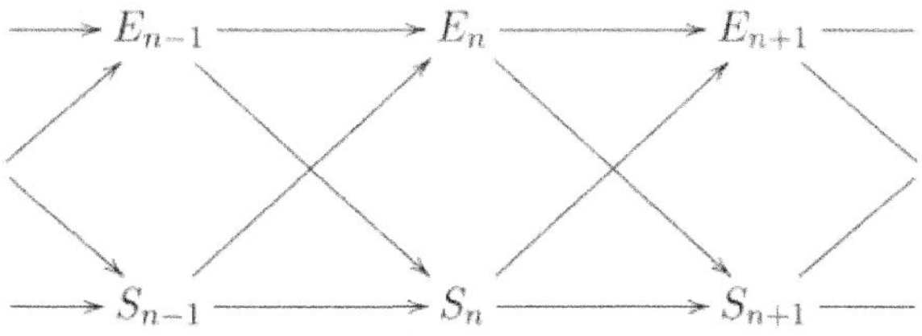

The fidelity with which an informational individual preserves its informational distinction over time defines an enduring entity relative to its entropic and fluctuating environment.[8] But maintaining that informational integrity requires successful "communication" between entity and environment, wherein information must also be faithfully conveyed.

Information and Meaning

Just because information is being conveyed between a system and its environment doesn't tell us anything about its "meaning," though. Shannon's formulation of information

dealt with informational quantification and accuracy, but remained entirely mute on how to assess whether information was actually *significant to* an entity, whether it was important or of value to some system in context. But that is precisely the sort of "meaning" we are interested in here. According to Shannon, I could send you an email containing billions of completely random numbers, and so long as you accurately received what I sent, you had gained a good amount of "information." Clearly, this is a very impoverished notion of information. It ignores the crucial question of what information is *about*, what it might be *for*, and thus its "semantic" significance.

Recently, however, complexity theorists have been able to fill in what's missing from this incomplete account. Specifically, they have identified how to articulate *meaning* within information theoretical terms. If entities are defined as stable transmissions of information through time, then continued, accurate transmission has *intrinsic significance to* that entity so in-formed. *Some information transmission has an existential import.* Likewise, when the information identifying an entity requires energy from the environment with which it is coupled to be sustained, then information about the environment must not only be accurate but helpful. That is, it must convey information about the environment in a manner that can be of use to sustaining the information that defines the entity.

These sorts of ideas can be framed in the language of information theory. Specifically, because systems exist in environments, the information between them is *correlated.* Information about the state of the system directly relates to information about the state of its environment, and vice versa

(cf. Figure 1.2). This coupling of information between system states and environmental states is called "mutual information" since it entails a statistical correlation between the two.

Now, as we have seen, maintaining information requires energy. For a system to resist entropy and maintain itself, it must acquire this needed energy from its environment. Information from the environment can be *relevant* to that end—or not. Indeed, there can be all sorts of "information" conveyed in Shannon's sense; what determines its true relevance, however, is adjudicated by how the mutual information of the system–environment relationship directly influences the system's continued existence. That is, at the most fundamental level, mutual information can be *meaningful* to the extent it causally impacts the specific existential requirements of a thing-in-context by helping it resist entropic breakdown. Only when information contributes to the capacity of something to continue existing in the face of entropy does it acquire *meaning to* that thing. This is what information can be *about.* When it is, that information is called *meaning.*

Complexity scientists Artemy Kolchinsky and David Wolpert have been key to developing information theory in this way, allowing us to account for this sort of meaning and relevance mathematically. In a 2018 paper,[9] they differentiate this specific kind of inherently meaningful information as true *semantic information.* "Shannon information theory," they write, "provides various measures of so-called syntactic information, which reflect the amount of statistical correlation between systems. By contrast, the concept of 'semantic information' refers to those correlations which carry significance or 'meaning' for a given system."[10] Kolchinsky and Wolpert thus

assess meaning as arising out of the system–environment relationship and so offer "a theory of semantic information that is based purely on the intrinsic dynamics of a system in a given environment."[11] *Information is meaningful for an entity to the degree that it assists in helping it resist entropy.* As they put it: "We define semantic information as the syntactic information that a physical system has about its environment which is causally necessary for the system to maintain its own existence."[12]

At the most basic level, then, environmental information causally tied to a system's informational persistence is a special kind of information that we can call meaning. Meaningless information, by contrast, offers nothing to an entity's continuation. Expending energy attending only to meaningless information about the environment extracts a true cost from an entity and is therefore existentially hazardous. If investments of energy into gaining information do not yield returns of energy commensurate with the expense, the entity risks entropic dissolution "Thus," Kolchinsky and Wolpert note, "there is a connection between 'paying attention to the right information', as measured by semantic efficiency, and being thermodynamically efficient."[13]

Meaning, they say, increases a system's *viability.* By "viability," though, they are not just referring to living systems. Kolchinsky and Wolpert's theory of semantic information is truly generalizable in that it "applies not only to living beings but also any physical system—whether a rock, a hurricane or a cell."[14] All things face entropy. The viability of some entities is simply more sensitive to environmental conditions than others, but all are susceptible to annihilation. Meaning, it seems, has a

truly expansive sense that can refer even to nonbiological entities—a point we shall repeatedly return to.

In sum, Kolchinsky and Wolpert's work, along with that of physicists like Carlo Rovelli,[15] reveal that information can be useful or useless, relevant or irrelevant, *about* something important or a harmful waste of time and energy. Information is relevant if it gives an entity causal power to maintain itself in its environment. Such information is then *meaningful* information, which emerges through the impinging of reality on agent behavior, which must adapt to its conditions to maintain its continued existence. Once we appreciate these fundamental dynamics about how systems relate to meaningful information in context, we can begin to understand the universal importance of *learning*.

Semantic information theory lies at the heart of neuroscientist and science writer Bobby Azarian's "integrated evolutionary synthesis," a key component of his ambitious Unifying Theory of Reality, which sees the whole of cosmic evolution as one great *learning process* unfolding across multiple levels of emergence. Linking theories like universal Darwinism, universal Bayesianism, evolutionary theory, and evolutionary epistemology into a single metatheory, Azarian's integrated evolutionary synthesis identifies the processing of semantic information as the basis of *knowledge creation* across all scales of evolution. Essentially, it is the driving logic of how all things persist, diversify, and grow. As he puts it in his 2022 book *The Romance of Reality: How the Universe Organizes Itself to Create Life, Consciousness, and Cosmic Complexity*:

> What is the foundation for a non-reductionist theory of everything? To understand the nature of life, and the nature of reality, we must begin with what many scientists consider to be the law of all physical laws, the second law of thermodynamics. …The natural tendency toward decay and disorder described by Boltzmann's statistical version of the second law—which can be summed up as "things fall apart"—gives any conceivable adaptive or sentient system an eternal existential challenge. Biologists call it survival, cyberneticists call it persistence, and physicists call it staying far from thermodynamic equilibrium. To stay far from equilibrium, in other words, to continue to exist, a system must be able to extract free energy…from a noisy, fluctuating, ever-changing environment. This is no trivial task. In fact, it absolutely requires that life *acquire information about the world it inhabits.* This information has intrinsic meaning to a system because it is causally necessary for its continued existence. Due to its inherent meaningfulness, this kind of information has been called …*adaptive information*… Because adaptive information reduces environmental uncertainty—or Shannon entropy, which is a technical term for ignorance—we also call it *knowledge.*[16]

As entities complexify, they become more adept at processing meaningful information from their environments. Expressed in concise terms: "life's thermodynamic imperative necessitates knowledge accumulation."[17] What Azarian calls "knowledge" here I will call *meaning* (or *meaningful knowledge*, when appropriate), following Kolchinsky, Wolpert, Rovelli, and

others. That is the sort of information that has true *relevance* and *significance* for an entity—knowledge with causal power on the continued viability and flourishing of systems in contexts.

And why, exactly? Some basic philosophical reflection on this idea reveals it to be both tautologically simple yet existentially profound. Without meaning, an entity would cease to exist, such that there would be nothing to speak of, nothing to consider. There would be no information of it at all, nothing to distinguish it, nothing to define it. If meaning is information that is relevant to the robust continuance of an entity, then the very fact that anything exists entails the existence of meaning. *For a thing to be entails that it has meaning.*[18]

From the very logic of semantic information such as Kolchinsky and Wolpert articulate it, it is clear that entities more effective at attending to meaningful information will be more viable and thus more evolutionarily successful. There is, then, a natural evolutionary incentive for entities to acquire increasing information about themselves and their environment in furtherance of this process. More information processing capacity, though, requires entities that are themselves more informationally rich. As this feedback cycle unfolds, entities *complexify*,[19] which both demands and occasions access to still *more energy* from the environment. And with energy increasingly needed to sustain such high-informational order in the face of entropy, the energy-information cycle perpetuates itself.

When we appreciate that 1) entities exist (i.e., faithfully convey their informational identity into future states) through correlation with the information of environmental contexts ("mutual information"), 2) this correlation is the basis of

meaning, and 3) entities are complexifying via this informational exchange, then we are led to an interesting conclusion: namely, that *meaning itself must be complexifying.*

Meaning and Complexification

If energy, information, and structural order are indeed all related, then we should expect to see a clear correlation between these measures in a complexifying universe. As entities move further from equilibrium, they should require more energy to sustain their informational richness and organizational depth. That is, we should expect to find greater energy flows per increasing measure of structural complexity. And, indeed, that is precisely what the calculations reveal.

In his book *Cosmic Evolution: The Rise of Complexity in Nature*, physicist Eric Chaisson charts the universe's trajectory towards greater complexity by mapping these interrelated measures across big history. In full accordance with the Second Law, he writes,

> the growing departure from thermodynamic equilibrium allows the cosmos to produce *increasing* amounts of net entropy *and* macroscopic information. We thereby have a means to appreciate in the main, if perhaps not yet understand the particulars, the observed rise in complexity throughout the eons of cosmic evolution.[20]

See Figures 1.3, 1.4, and 1.5 for his graphs of these trends.

Figure 1.3. Departure from equilibrium

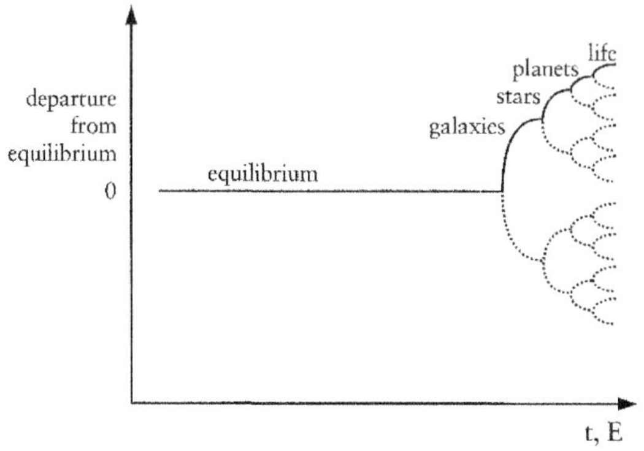

Figure 1.4. Energy rate density

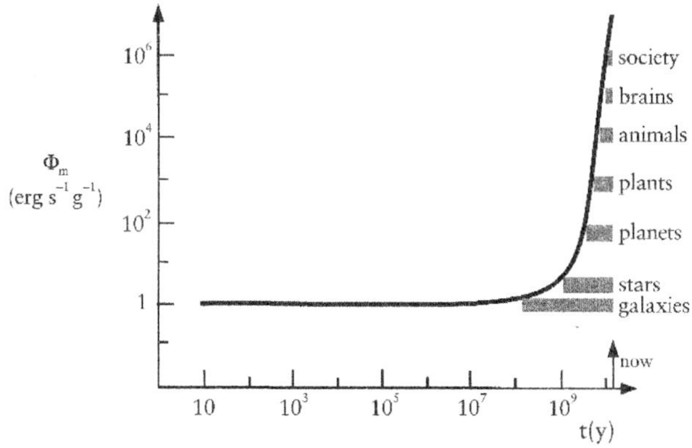

Figure 1.5. Information[21]

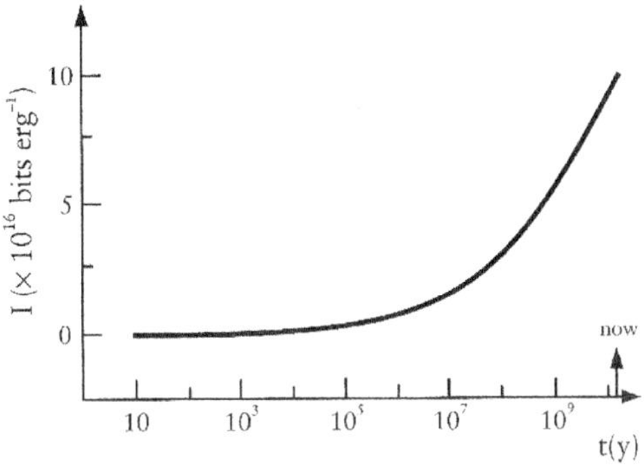

From an initial state of homogeneity soon after the Big Bang, the universe began to self-organize and complexify. Chaisson's graphs all show a cosmos moving further from equilibrium, with entities of increasing free energy rate density and informational richness over time.

Indeed, using free energy rate density (Φ_m) calculations alone,[22] Chaisson charts a clear trend towards complexification from matter to plant life to animals to culture. Citing some key figures in a 2014 article, he summarizes:

> Among physical systems, stars and galaxies generally have energy rate densities (10^{-3}–10^2 erg/s/g) that are among the lowest of known organized structures. …For plant life on Earth, energy rate densities are well higher than those of normal stars and typical galaxies… Likewise, as animals evolved from fish and amphibians to reptiles, mammals, and birds, their Φ_m values rose still more. …Not surprisingly, brains have among the highest values of Φ_m for all living things. …Social evolution can be tracked, again in terms of normalized energy consumption, for a variety of human-related cultural advances among our ancestral forebears, from early agriculturists ($\sim10^5$ erg/s/g) to modern technologists ($\sim10^{6.5}$).[23]

Figure 1.6 provides a visual representation of this complexification trajectory.

Such energy density increases correspond to increasing informational richness. Indeed, Chaisson writes, "to us, information basically *is* a form of energy."[24] Thus, as the universe evolves more complex entities, more and more

information is being generated—*including the semantic information for those entities in context.* Because increasing energy density and structural complexity are cause for ever increasing semantic information processing, Azarian calls cosmic complexification itself a *learning process*:

> By now, we can see a simple story emerging. By reducing Shannon entropy, or ignorance, life is able to extract the energy it needs to reduce Boltzmann entropy, or disorder. The growth of knowledge and the spread of organized complexity, therefore, go hand in hand. The second law is the impetus for learning.[25]

Figure 1.6. Complexification and Energy Rate Density[26]

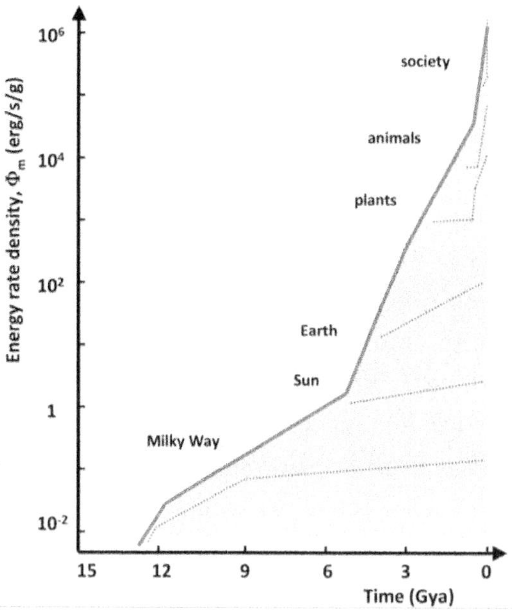

Learning thus appears to be the natural consequence (and driver) of a complexifying universe, as more energetically dense entities emerge to embody and encode more information about their environments.[27]

In this process, the very *means* by which information gets processed complexify, with novel computational mechanisms emerging out of earlier ones to process new kinds of information. As physicist and quantum computing theorist Seth Lloyd expresses it:

> The history of the universe in terms of information-processing revolutions, each arising naturally from the previous one, already hints at why a computing universe necessarily gives rise to complexity. In fact, we can prove mathematically that a universe that computes must, with high probability, give rise to a stream of ever-more complex structures.[28]

If increasingly complex entities-in-context are processing more and more semantic information to maintain themselves in the face of entropy, then we must conclude that *meaning itself has been complexifying through time.*

But what does this actually mean? Is meaning really growing more intricate, refined, nuanced, rich, deep, sophisticated, etc.? What does highly complex meaning look like compared to less complex meaning? To unpack this idea, we must examine these successive "information-processing revolutions" Lloyd refers to which have punctuated the complexification of the universe, such as marks the transition from matter to life. We must follow the arc of cosmic

complexification and identify just what entities are processing what sorts of meaningful knowledge in a manner that maximizes their viability, and how this leads to qualitatively deeper forms of meaning.

The Unified Theory of Knowledge (UTOK) is a big history framework that offers perhaps the best map of cosmic complexification to date. It was developed by Gregg Henriques, a professor of psychology at James Madison University, with the intent to properly situate psychological phenomena within the context of the natural sciences and evolutionary history. In the process, it has helped clarify not only the field of psychology but emergent evolutionary history more broadly, presenting a convincing metatheoretical synthesis of different scientific disciplines that all tell one coherent narrative of cosmic complexification.

Two components of the framework specifically—the "Tree of Knowledge" (ToK) diagram and the "Periodic Table of Behavior" (PTB) chart—bring especial clarity to the way information processing by thermodynamic entities in their respective environments has led to a universe of stratified behavioral patterns constituting genuinely novel planes of existence. As Henriques notes in his 2023 book on the UTOK framework, *A New Synthesis for Solving the Problem of Psychology*:

> The emergence of information theory, information science, chaos and complexity sciences, and philosophical analyses of living and mental creatures…all show why a strong physical reductionism or eliminative materialism is a fallacious frame for scientific knowledge. The ToK and PTB show how the natural world is singular and monistic

> in the sense that the foundational ground of existence is an Energy Information Field, out of which the dimensions of behavioral complexification called Matter, Life, Mind, and Culture emerge. …The latter three all give rise to complex adaptive dynamic systems, whose patterns of behavior are not just matter and energy on a field of space and time but include dynamically self-organizing complex adaptive systems that engage in information processing and communication behaviors that require different metaphysical concepts above and beyond efficient causation.[29]

That is, contrary to older reductionist notions long prevailing in the sciences, phenomena at the levels of Life, Mind, and Culture are *emergent*, exhibiting behavior inexplicable by a reductionistic analysis of material parts in isolation.[30]

These emergent planes are such because of the novel kinds of information processed at each successive level. As Henriques summarizes it:

> According to the ToK system, *genetic* information processing gives rise to the dimension of Life, *neuronal* information processing gives rise to the dimension of Mind, and *symbolic* information processing gives rise to the dimension of Culture.[31]

More specifically, I will argue, we can actually say that *these are novel kinds of meaning that have emerged*, since these are kinds of information that entities may process in the learning of knowledge that bears on their viability. Genetic meaning

speaks a different language than neuronal meaning, which speaks a different language still from symbolic meaning.

These novel information processing systems create unique adaptive contexts that must be navigated according to their own informational demands, such that meaningful information has new emergent properties at each new level. The Tree of Knowledge diagram maps these levels, showing the emergence of Life from Matter, Mind from Life, and Culture from Mind (Figure 1.7).

Given these unique informational contexts, each level thus sees its own *unique kinds of system–environment couplings*. The UTOK's Periodic Table of Behavior (PTB) chart provides a concise map of these increasingly complex kinds of "entity–field relationships" that arise within the different complexity planes of Matter, Life, Mind, and Culture (Table 1.1).

Figure 1.7. The UTOK Tree of Knowledge[32]

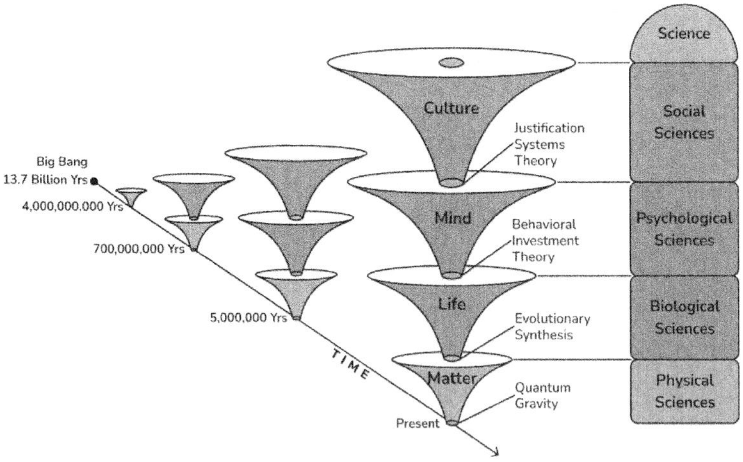

Table 1.1. The UTOK Periodic Table of Behavior[33]

Plane of Existence		Matter	Life	Mind	Culture
Information Type			Genetic	Neuronal	Symbolic
Entity–Field Relationship	Behavioral Entity	*Object*	*Organism*	*Animal*	*Person*
	Context of Behavior	*Field*	*Ecology*	*Environment*	*Society*
Primary Levels of Object Analysis (Part, Whole, Group)	Groups of Wholes	Molecule	Multicell/ Colony	Family/ Group	Community/ Nation
	Primary Whole	Atom	Cell	Mind^1	Mind^3
	Primary Part	Particle	Gene	Neural Network	Symbolic Justification

As Henriques summarizes it: "The PTB affords us a new epistemological lens that creates an ontological taxonomy of entity–field relations across the levels and dimensions of analysis that make up the stratification of behavioral complexification in nature."[34] These develop across material, biological, psychological, and cultural levels of complexity.

Thus, we see that the entity–field relationship unfolding through the four planes of complexification can be identified as follows:

1) the *object–field relationship* of the Matter plane,
2) the *organism–ecology relationship* of the Life plane,
3) the *animal–environment relationship* of the Mind plane, and
4) the *person–society relationship* of the Culture plane.

Each coupling, I posit, creates mutual information correlating system with environment. Because of that, each coupling represents unique demands on viability, and thus a domain where unique qualities of meaning manifest.

In fact, each entity up the complexity stack must engage with *all* of the various behavioral contexts below it. That is, for example: humans must navigate their socio-ecological culture, yes, but also the ecologies of animals, the environments of organisms, the fields of objects, and all of the permutations of their interactions. A human must have *meaningful relationships* with all these domains if they are to continue existing effectively in the world.

Such relationships are not meaningful only in subjective ways, but rather in a manner that is intricately tied up with reality going all the way down to thermodynamic necessity. To misinterpret the meaning of a grizzly bear (animal–environment), or *E. coli* (organism–ecology), or a steep cliff (object–field) is to face entropic annihilation. To say it again, *meaning* of this kind has causal power; to meaningfully know a thing is to integrate its information in an adaptive way. Meaning is what enhances viability at every level by informing the behavioral interventions needed for entities to thrive.

Thus, the more complex an entity is, the more complex information it must process across the numerous planes of reality. While all semantic information does ultimately "cash out" at the thermodynamic level, there are, it must be repeatedly stressed, novel layers to meaning introduced as complexity unfolds. These novel layers emerge as more complex information is processed through the universal learning process, such that meaningful information gets generated differently as you move up the complexity stack but accumulates in an accretive way.[35] Meaning for cells is processed as genetic information; but meaning for an animal modifies mere genetic selection with an overlay of neuronally

meaningful information processing. For humans, viability-enhancing meaning must be sought across both of these informational domains as well as through symbolic exchanges.

Complexification and Learning

If it is true that meaning arises from the mutual information correlating entities and environments, and that these entity–field dynamics complexify through the emergent levels of Matter, Life, Mind, and Culture, then entities at *all* levels can be understood as engaging in an increasingly complex learning process whereby more sophisticated forms of meaning emerge. If for anything to exist means it has navigated its environment successfully enough to have not succumbed to entropy, then there is meaningful information linking that entity to its conditions, and so what Azarian calls "knowledge"—even at the most basic level.

The learning process of entities in the Universe, then, can be understood as extending from the most complex adaptive organisms like human beings on the Culture plane, down to animals in their environments on the Mind plane, down still further to biological entities like single-celled organisms and plants on the Life plane, and, I would argue, even shades down into the plane of Matter, where inanimate objects can complexify through pre-Darwinian forms of evolution (e.g., autocatalysis, dissipative adaptation) and whose very existence as objects may even arise from quantum decoherence mechanisms based on similar selection pressures (though

calling this "learning" obviously stretches the term to its breaking).

To delineate these distinct kinds of learning relative to complexity level, and thus their relative kinds of meaning, we can speak of:

1) Structural Learning (Matter),
2) Genetic Learning (Life),
3) Cognitive Learning (Mind), and
4) Symbolic Learning (Culture).

Through these increasingly complex forms of learning, purpose and value emerge to grow more rich, comprehensive, and integrated.

The majority of this book series will focus on the sort of learning that takes place on the Culture plane. But before we arrive there, let's take a journey through the preceding levels that lead up and give rise to such complex human meaning: a path connecting us to the universal learning process.

1. Structural Learning (Matter)

We start at the lowest level, begun nearly 14 billion years ago, where "learning" seems, at best, a semantically loose description for the sorts of processes underway. That is what we should expect, though, if we mean to trace the continuum from the crudest, most provisional forms of adaptive information processing to the most complex. At this basal level, we are dealing with the very rudiments of the objective

world and how they come into being. Here we ask: Could this process also be subject to the sort of viability-filtering mechanisms that characterize genuine learning at higher levels of complexity?

John Campbell thinks so. In his books *Universal Darwinism: The Path of Knowledge* and *The Knowing Universe*, Campbell draws on the 2009 work of theoretical physicist Wojciech Zurek of the Los Alamos National Laboratory[36] to show how quantum decoherence (the process by which quantum states become determined as classical reality) operates in terms of inferential information processing. Since the early 20th century, quantum mechanics has puzzled physicists by its fundamentally probabilistic nature, which is so unlike the deterministic certainty of classical physics. Do *all possible states* of quantum phenomena exist then somewhere—somehow—perhaps in some sort of "multiverse"? The seeming gulf between a quantum world of probability and a classical objective world of definite objects has left theoretically unresolved the transition from one realm to the other. Zurek, Campbell, and other proponents of "quantum Darwinism" resolve this question by seeing the transition to the classical objective world as one of information transfer subject to environmental selection. "[O]nly the effects of quantum systems that can pass through the filter of decoherence compose our reality," Campbell writes.

> Quantum systems participate in reality only through those interactions. When analysed in detail these interactions or quantum decoherences consist of a transfer of information between the quantum system and the web of

> objective reality. Not all information concerning the quantum system is transferable. In fact the vast majority is not transferable. The relatively tiny amount of information that can be transferred is selected from the huge range of quantum possibilities and numerous copies of this information are deposited in the environment by a Darwinian process…[37]

In this way, when interactions occur, only the environmentally viable information is selected from the vast variability of possible states. The resulting conditions then form the informational basis for determining future selection, and so on.[38] Because of this, the very fabric of reality arises through a sort of "quantum computation," which is continually updating the realizable possibilities of future states of the universe. So framed, the objective universe is itself the acquired "knowledge" (i.e., adaptive information) of a quantum learning process.[39]

Such quantum computational models are compelling, but admittedly speculative. Once entities have entered the classical plane of deterministic Matter, however, we are on surer ground. It is clear to see how interactions among physical particles unfold within the "object–field" relationship of the PTB. For a particle's activities become impinged upon by other particles and forces in its environment, which affect its viability as a stable, persisting entity. In the proto-Darwinian struggle for survival—not of the *fittest* but simply of the *stable*—particles that formed into enduring atoms out-competed particles that did not, and so persisted. Amongst the vast variation, some stability gets selected for by the environment. And though

material objects are not in any way animate agents in this process—let alone conscious or volitional ones—the end result is the endurance of structure against entropy through being well-fitted to environmental niche: the basis of the entity–field relationship and so also of meaningful information.[40]

But, again, is this really "learning"? Admittedly, given the very minimal degree of complexity at this level, using a word like "learning" for such interactions, which are so alien to our sense of the term, may seem specious. Nevertheless, it is the indisputable *continuity with later forms of learning* that, I believe, renders such an interpretation eminently valid.

Indeed, one need not rise all the way to the level of Life to see this made plain. The "dissipative structures" described by Nobel-winning chemist Ilya Prigogine represent a much clearer instance of purely material objects navigating the energy dynamics of their environment to push back against entropy and maintain structural persistence—the essence of learning "semantic information" as defined above. Such structures include entities like whirlpools, hurricanes, and tornados, which more effectively dissipate free energy through the spontaneous emergence of greater order.[41] Indeed, we can recognize that they arise, functionally speaking, *to* dissipate that energy.[42] If meaningful information is what allows an entity to maintain its informational structure in the face of entropy, then we are forced to conclude that such entities even have meaning. From here, one can track the process into genuine "dissipative *adaptation*" such as Jeremy England and others are exploring, whereby the specific environmental conditions of the energy itself actually *select for* what structures emerge.[43]

Teleological or "goal-directed" behavior is thus evident even in prebiotic entities. That is, all of this learning unfolds entirely on the plane of Matter. And while the details of abiogenesis out of dissipative adaptation remain to be filled in, there are promising signs that the evident continuity between material dissipative structures and biological organisms (which are clearly a subset of the former) will offer new revelations about the origins of Life. At the very least, it shows us that an entity need not be a full-blown biological organism to exist in an entity–field relationship whose correlation defines meaningful knowledge to the entity. Rather, a very crude, simplistic form of meaning exists even for a whirlpool (as Kolchinsky and Wolpert's framework posits).[44]

While this may seem at first to diminish the very idea of meaning, it actually does the opposite, I believe. For, by showing that meaning connects down into the very rudiments of reality, we can see the error (common these days) of presuming it to be merely subjective, or as a thing only humans "make up" as some arbitrary "social construction." To the contrary, meaning, along with a related word, "teleology," are both quite real—something cyberneticists and complexity scientists have long appreciated as they have endeavored to explain goal-directed material phenomena like dissipative structures. As Azarian concludes, with dissipative structures

> a natural pressure to minimize free energy and produce entropy spontaneously creates an ordered structure with an objective function. With the emergence of dissipative structures, we see the emergence of purpose in nature...[45]

2. Genetic Learning (Life)

Of course, the universal learning process (and thus meaning) becomes appreciably clearer to recognize when, around 3.8 billion years ago, we see its first truly sublime product: Life—a particular kind of dissipative structure that can also *reproduce itself via meaningful information.* Such is the system–environment knowledge contained in the "genetic code." Here, meaning truly comes into view, as the goal-directed character of biological behavior becomes overt and undeniable.

On this point, the work of neuroanthropologist Terence Deacon has been especially elucidating, offering a profound contribution to the reappraisal of meaning and teleology in the biological sciences. Since Darwin, teleology had been summarily eschewed from naturalistic descriptions of life and its development, replaced instead by an emphasis on blind meandering and undirected chance. Having identified the crucial *efficient cause* of natural selection (that is, the basic material cause-effect mechanism), biologists of the 19th and early 20th centuries believed they could then dispense with the idea of any teleological *final cause* such as Aristotle had once identified for organisms—which is to say, the idea that their end, goal, or purpose structurally *in-forms* them.

Ultimately, however, the failures of a purely reductionistic biology to adequately explain biological organisms without reference to goals and aims has led to a paradigm shift in the field, as emergentist frameworks from complexity science have helped shed important new light on long-vexing problems about biological behavior. Deacon's work is a significant

contribution to this paradigm shift. In his comprehensive 2012 book *Incomplete Nature: How Mind Emerged from Matter*, he offers a scientifically robust and philosophically compelling reconsideration of teleology's role in biology. By emphasizing the inherent lack or need of goal-directed entities (i.e., their "incompleteness"), Deacon removes teleology from a speculative metaphysical register and places it comfortably within a processual physical one. This allows us to see the commonalities that exist between all entities intrinsically constituted in processual relation to an absent goal state. As he puts it:

> Such concepts as information, function, purpose, meaning, intention, significance, consciousness, and value are intrinsically defined by their fundamental incompleteness. They exist only in relation to something they are not.[46]

Such absence thus "motivates" (in the most basic etymological sense, we might say: "causes to move, *animates*") the behavior of biological entities.

This dynamic is clearest in highly complex organisms like ourselves, who *aspire*, *wish*, *will*, *intend*, etc. Meaning and purpose are clear, we might say, within the PTB's "person–society field" of Culture. But appreciating the deeper dynamic of motivational absence allows us to trace the quality of such meaning back down through less complex forms—which is precisely what we have been endeavoring to do here. As Deacon proposes:

> Because it is complex, the concept of mental purpose can be progressively decomposed, revealing weaker forms of this consequence-organized relationship that do not assume intrinsic mentality. Thus the concepts of function, information, and value have counterparts in living processes, which do not entail psychological states, and from which these more developed forms likely derive.[47]

To distinguish from the associations of purpose with more complex, cognitive versions of intentional behavior, Deacon coins the term "ententional," which can apply also to unconscious goal-directedness at the purely biological level. Ententional entities are teleologically driven. As Deacon puts it explicitly: "Ententional processes and relationships can best be classed as expressions of final causality in Aristotle's terms."[48] In short, *living organisms have meanings.*

But unlike in ancient theological or speculative metaphysics, such meaning is not an essence directly bestowed upon creatures by a transcendent Creator, but rather a product of the intrinsic relational dynamics between an organism and its ecological context—that is, we might say, it *emerges from the semantic information linking a biological organism with its ecological niche.* As Deacon puts it:

> [E]ntentional processes have a distinctive and characteristic dynamical circularity, and…their causal power is not located in any ultimate stuff but in this dynamical organization itself. Our ultimate scientific challenge is to precisely characterize this geometry of dynamical forms which leads from thermodynamic

> processes to living and mental processes, and to explain their dependency relationships with respect to one another. It is a quest to naturalize teleology and its kin…[49]

In thermodynamic terms, the telos of a biological organism is, like other dissipative structures, to remain far from equilibrium.[50] Behavior directed towards this goal (e.g., a bacterium navigating towards nutrients through chemotaxis) is purposive behavior, and information that assists in this purposive behavior is meaningful information.[51]

What distinguishes the meaningful knowledge unique to Life, however, is a novel information processing language: genetics. This, according to the UTOK Tree of Knowledge, is what marks the emergence of a novel type of entity exhibiting a novel form of behavior. Whereas dissipative structures arise spontaneously and then disappear as conditions demand, Life is different in that it has been able to persist and reproduce without interruption since its initial emergence some 3.8 billion years ago. *By being able to store meaningful information in the genetic code and transmit this information through reproduction, an entirely new level of complexity has taken hold in the universe.*

Specifically, Life stores meaningful information *about the environment* in DNA. With this, we begin to see an early form of representation in the cosmic learning process. Deacon stresses this *aboutness* as a crucial characteristic of the mutual information linking organism and ecology, writing:

> Living organisms are far-from-equilibrium systems that are dependent on transforming the physics of their interaction with the environment into information about

> that environment. …This is why organism features reflect the constraints of their environments, and what justifies claiming that the information embodied in a phenotype and its genotype is information *about* this environment. But this *aboutness* only reflects these environmental constraints *with respect to* the constraints of the formative processes critical to reproduction and self-maintenance of that organism.[52]

Meaningful information for biological organisms encodes relevant referents to the world that the organism inhabits. Such representations, however, are not stored in the memory of the individual organism—that does not arrive until the level of Mind, when nervous systems/brains emerge. Rather, *these representations are translated into the language of DNA itself.* The very information that shapes the organism includes semantic information about its world. Organisms *are* embodied meaning.[53]

Crucially, though, life does more than store information about its environment—it also *updates* that information. Through the natural selection of evolution, meaningful information gets filtered according to environmental pressures, creating, over time, an increasingly intimate coupling between organism and ecology. The meaning inherited by an organism in its DNA is matched against its environmental conditions. As this "signal" gets transferred across generations, it is subject to entropy just like any kind of Shannon information. This introduces variation in the signal, and these variant meanings can be better or worse adapted to their conditions. Better adapted informational couplings are selected and get carried

forward, such that meaning comes to carry more significant information about the environment. In this way, genetic evolution and natural selection add a crucial pruning mechanism to the transmission of meaningful information through time.[54] As Deacon puts it in his essay "Shannon—Boltzmann—Darwin: Redefining Information":

> What counts as useful information in biological evolution is determined *after the fact* with respect to its ability to pass through the functional error-correction mechanism of natural selection. The "accuracy" of the inheritance signal is both "tested" and refined by the way the far-from-equilibrium dynamics that constitutes organism development, maintenance, and reproduction conform to environmental constraints and opportunities. Evolution is thus a generator of information for the organism and a process that rectifies this information with its reference. Thus, although the evolutionary process is itself non-normative, it produces organisms capable of making normative assessments of the information they receive.[55]

In sum, evolution through natural selection behaves as a learning process at the level of Life. It is the special mechanism at this level that allows meaningful knowledge to be stored and updated, furthering the complexification of energy, information, and structure.[56]

Deacon's mention of normativity above highlights a crucial point, however, and one that will become increasingly important as we move forward. That is, nature's learning process we have been describing produces knowledge that is

meaningful only to the degree that it assists the entities producing it to better navigate their entity–field relationship. That is to say, information gets its causal power to the degree it is *of value* to the existence of the entity—i.e., whether it assists the entity in its goal to remain far from equilibrium. *Value*, then, is also an inherent product of the entity–field relationship. Value does not exist "in and of itself," but in the dynamic adaptive contexts relating organism to ecology, animal to environment, etc.—that is, in the process of energy informing structures adapted to specific environments.

Another way to express this is to say that value ("good vs. bad") must be understood as deriving from the entity–field relationship, wherein there is a relative "good *for*" an entity, as well as a relative "bad *for*" that entity. For a bacterium on the Life plane, navigating via chemotaxis towards a collection of nutrients is *good for* its continued existence, aiding its efforts to obtain free energy to maintain itself in the face of entropy and so persist. Conversely, navigating into an area flooded by toxins is *bad for* the bacterium, threatening its basic integrity as an organism. Even though the single-celled organism is among the most primitive of entities on the Life plane, we can still identify *values for* the organism, which have *meaning for* it as information with causal power relative to its flourishing or annihilation.[57]

In fact, Deacon himself recognizes that this normativity actually goes even *further back than Life.* He points to the example of autocatalysis, a chemical process whereby a collection of molecules reciprocally and recursively create more and more of themselves. This feedback loop is an ententional process in the sense that it is goal-directed and thus

purposive in its behavior. More complex than a whirlpool, but still simpler than a full-blown living organism, it likewise has its form of normative "value." As Deacon recognizes:

> Functions are normative to the extent that they fail or succeed to achieve some end. ...So, even these simple molecular systems have crossed a threshold in which we can say that a very basic form of value has emerged, because we can describe each of the component autogenic processes as there for the sake of the autogen integrity, or for the maintenance of that particular form of autogenicity. Likewise, we can describe different features of the surrounding molecular environment as "beneficial" or "harmful" in the same sense that we would apply these assessments to microorganisms. More important, these are not merely glosses provided by a human observer, but intrinsic and functionally relevant features of the consequence-organized nature of the autogen itself.[58]

Surprising as it sounds, because meaning exists even on the level of Matter, there is "value" even for purely material entities to the extent they resist entropy and are thus goal-directed. Like meaning itself, this does not diminish the idea but rather makes it all the more theoretically robust. Meaning and value exist, not as mere social constructs or subjective inventions, but as *real* aspects of the world. They do not, we must stress, inhere in things themselves, as essences bestowed (such as traditional theological thinking proposed); but neither are they purely subjective experiences overlaid onto the world (such as modern reductionistic thinking has tended to assume).

Rather, they arise in the dynamic relationship *between* an entity in its contextual field. Meaning is real *because* it is contextual. Meaning, you could say, is a *real relationship*.[59]

In terms of the UTOK map of complexification, the Life plane is defined by meaning processed as genetic information. Indeed, as many biologists have argued, the actual informational individual here is not so much the single organism as the broader gene pool—the species. Viability is adjusted for the group, not just the entity, so *meaning transcends the organism*. At the genetic level, meaning is fully collective, since learning is only collective at this stage.[60]

With time, though, through the success of viable species, still more complex information processing systems evolved. Specifically, a certain kind of multicellularity led to the emergence of neuronal nets that developed into increasingly complex nervous systems—an integration of information that gave rise to an entirely new emergent dimension of complexity: Mind.

3. Cognitive Learning (Mind)

The proliferation of networked systems of nerve cells (i.e., nervous systems) with a central processing center (i.e., brains) that we see marking the "Cambrian explosion" (c. 530 million years ago) occasioned the emergence of a novel dimension of complexity and information processing. At this new level of "minded animals" (Mind), the PTB entity–field relationship is that of the animal–environment coupling, and with such animals we see the emergence of *neurocognitive computation* of

meaningful information. Meaning becomes sensory, mobile, and, eventually, *subjectively experiential.*

The evolution of mobile bodies that could freely navigate the environment meant organisms could now learn in the sense more familiar to us—that is, by direct sensory engagement with the environment in real time. Before, at the level of Life, learning occurred only at the population (phylogenetic) level, as natural selection "updated" the species' genetic code to better match environmental conditions. Meaning was passed down through genetic inheritance from generation to generation. It was embodied, but not experienced; reflexive, but not reflective. Memory was something in the genes, not in the head.

With the emergence of nervous systems, however, this changed. For one thing, it saw the evolution of heads! (Natural selection hit on the bilateral symmetry of the animal body plan, with the nervous system's central processor, its brain, positioned in the anterior.) With brains, the updating of meaningful information could now take place at the individual (ontogenetic) level, as individual organisms could now revise their own internal models of the world based on sense inputs from their surroundings in real time.[61] Such information could then be stored in individual cognitive memory akin to the way information had been stored as collective genetic memory. This meant that, in addition to the relatively slow pace of genetic adaptation, there could now be much quicker behavioral adaptation.[62]

All this entailed a massive upgrade for the universal learning process, allowing information of much greater complexity to be processed at much greater speed by evolving

animals. It also entailed that animals develop increasingly sophisticated *internal models* of the world they inhabited. As Richard Dawkins has put it, "Survival machines that can simulate the future are one jump ahead of survival machines who can only learn on the basis of overt trial and error. The trouble with overt trial is that it takes time and energy. The trouble with overt error is that it is often fatal. Simulation is both safer and faster."[63] At the complexity level of Mind, animals learn to simulate the world with greater and greater reliability in their bid to reduce ignorance about the environment and remain far from equilibrium.

This is the core insight of Karl Friston's theory of cognition, Active Inference, which is based on the "free energy principle" of predictive processing.[64] According to this "unified brain theory," the mind is a complex adaptive system that allows animals to stay far from equilibrium specifically by modelling their environment and continually updating that mental model through comparisons with how well actual sensory inputs match its expectations. Based on these calibrations, animals adapt their behavior and get better at predicting the world—AKA, *learning*.[65]

To achieve this, however, renders meaning into a mental computational challenge. What information about the world, exactly, should be included in the nervous system's mental model? What can be safely ignored? There is, we have seen, an infinite amount of "information" in Shannon's sense that might be considered. How does a cognitive entity single out only the *relevant* subset of that information to optimally navigate the environment? How can neurons detect viability? How can *significance* be computed?

John Vervaeke, a 4E cognitive scientist and professor of psychology at the University of Toronto, has proposed a theory of "*relevance realization*" that addresses this issue and extends the sort of meaning we have been considering into the register of Mind. The four E's of 4E cognitive science refer to the *embodied*, *embedded*, *enactive*, and *extended* nature of mind; that is, cognition is 1) localized in embodied organisms, 2) embedded in environmental contexts, 3) enactive of the world with which it engages, and 4) able to be extended beyond the internal bounds of the brain through technologies like tools and writing.[66] Such a framework has proven crucial for moving the field of cognitive science beyond simplistic input-output computational models by appreciating the mind's fundamentally incarnated and contextualized emergence from complex adaptive biological systems in rich ecological contexts. All of this is necessary for appreciating how minded animals process information as *meaningful.*

Consonant with our previous considerations of meaning, Vervaeke's theory of relevance realization emphasizes the coupling of system and environment. He refers to the entity–field relationship at this level as the "agent–arena relationship," the domain of agential subjects operating in objective contexts. Appreciating the crucial insight that meaning arises in relational exchange *between* subjects and their environment, Vervaeke coins the term "transjective" to capture the mutually-informing aspect of the subjective/objective agent–arena relationship. *Meaning*, he says, *is transjective*, arising from the dynamic interplay of subjects with(in) their objective world. This idea of "transjectivity" is helpful, and I will use it going forward to refer to the specific dynamic inherent to *all* entity–

field relationships, as it captures a key reality of meaning at every level of complexity.

For embodied, mobile agents navigating environments to achieve their goals, the thermodynamic imperative manifests as solving various situational problems (e.g., acquiring food, avoiding predators, etc.). To meet these challenges, says Vervaeke, agents must learn to accurately identify the *relevant* information in their "salience landscape." To this end, minded animals have evolved neurocognitive mechanisms that assist in relevance realization—to filter out probably unmeaningful information and home in on probably meaningful information. As Vervaeke puts it, "for an agent to take action in even relatively simple circumstances it must somehow intelligently ignore a great deal of information."[67]

This idea has clear connections to Kolchinsky and Wolpert's more basic formulation of meaning considered at the outset. As they articulated it, "there is a connection between 'paying attention to the right information', as measured by semantic efficiency, and being thermodynamically efficient."[68] Meaning is only a subset of all possible information. Vervaeke's relevance realization would model how minded animals learn to "pay attention to the right information." Indeed, so fundamental does this idea appear to virtually all aspects of cognition that he declares, "relevance realization is the criterion of the cognitive."[69]

As for *how* animal minds do this, Vervaeke and colleagues posit a number of distinct trade-off relationships built into the response dynamics of cognitive agents.[70] All of these represent so-called "opponent processes" of various types: contrasting orientations which ultimately converge towards a dynamic

equilibrium with the environment through ongoing recalibration in light of new sensory feedback.[71] That is, by continually negotiating the proper contextually-appropriate balance between compression (i.e., generalization) on the one hand vs. particularization on the other, assimilation vs. accommodation, integration vs. differentiation, animals find the "optimal grip" required by their context for achieving their specific goals (Figure 1.8). They write:

> Compression results in the assimilation of information to existing structures and therefore results in the integration of information. Particularization results in the accommodation of existing structures to information so that the differentiation of information occurs. Since the brain is doing both in an opponent fashion, it is simultaneously developing both integration and differentiation. A system that simultaneously integrates and differentiates its functions is complexifying as a system… This is important because complex systems have emergent functions. This is the way in which the brain can develop its competence to deal with a complex world, viz., it self-complexifies (or develops).[72]

In short, *Cognitive Learning is a complexification process*—a finding that shouldn't surprise in light of our considerations so far.

Figure 1.8. Relevance Realization and Complexification[73]

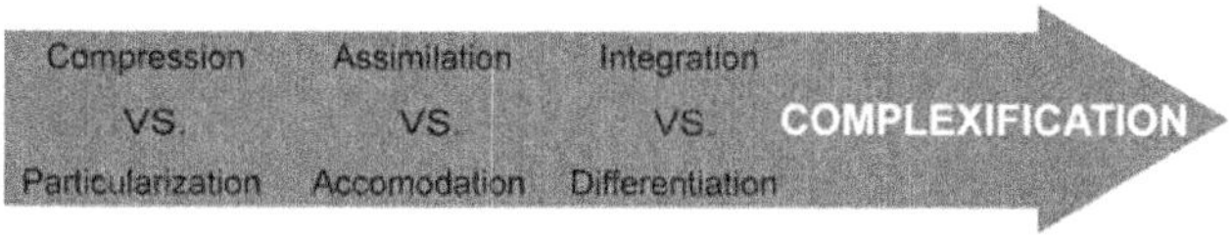

Such cognitive complexification amounts to a dynamic *developmental* process. As Vervaeke puts it elsewhere: "[C]ognition is inherently developmental in nature, rather than development just being the peripheral issue of how cognition emerges. Thus, we may speculate that developmental psychology ought be seen as central to cognitive psychology."[74]

Indeed, as we shall see in the following book, a dynamic "equilibrium" achieved via continual "assimilation vs. accommodation" was precisely the mechanism of human learning articulated by the constructivist developmental psychologists in the early 20th century. We can thus already anticipate how these same cognitive dynamics carry forward into the Culture plane, where Symbolic Learning progresses by means of hierarchical complexification of linguistic-semantic meanings, not just sensorimotor and perceptual schemas.

More immediately, though, we must finish our analysis of how cognitive meaning unfolds as the animal mind complexifies. Here we encounter some profound emergent properties of utmost importance. For, while we have been considering cognition and intelligence as they relate to Mind, one crucial concept has so far remained conspicuously absent from our discussion—one which is absolutely essential to all notions of meaning and value such as we as humans are familiar with.

Consciousness.

With consciousness, meaning becomes *experiential* for the first time. Meaning becomes subjective. It becomes, that is, what we as humans intuitively think of when we describe anything as *meaningful* or *significant* in some way. It becomes *felt* as meaningful.

But how did consciousness emerge out of unconscious cognitive information processing? In their 2019 book *The Evolution of the Sensitive Soul: Learning and the Origins of Consciousness*, Simona Ginsburg and Eva Jablonka trace this evolution in some detail. Consciousness, they posit, is a direct product of the neuronal infrastructure laid down by animals as they developed more comprehensive, integrated world modelling capacities. Cognitive *Learning*, they say, was what led to subjectivity. They write:

> When we think about consciousness and sentience, we do not usually think about learning. What comes to mind are the immediate experiences of seeing a sunflower, of feeling a stab of pain or a flood of joy, of hearing the song of a blackbird, of smelling freshly baked bread, of tasting a ripe banana. These experiences can be fleeting, and we may soon forget them. But how did animals come, evolutionarily and developmentally, to have such experiences? …[L]earning was the basis on which this experiencing process was constructed. There is no point in sensory discrimination in a rapidly yet recurrently changing world if we are unable to remember it and use it to improve our lot. Moreover, we suggest that even fleeting experiences involve processes of stabilization and updating that are intimately related to learning, as is most obvious during early development. …All pattern-perception and compound-action selection depend on learning because all rely on the (learned) construction of updatable models of the world in which the animal acts.[75]

Specifically, Ginsburg and Jablonka see subjective consciousness emerging in animals as they transitioned from simpler learning methods like reflex-based behavioral conditioning (or "associative learning") to the full repertoire of what they call "unlimited associative learning," which saw broad hierarchical integration of such responses along with a developed memory system that could store learned information in a temporally continuous manner.

One way to summarize this development in a way that brings to the fore our present concern is to say that *multiple domains of cognitive meaning became integrated* in animals, such that distinct purpose and value systems "synced up" in a holistic manner so as to better guide the organism's efforts to stay far from equilibrium. Ginsburg and Jablonka speak of a single, integrated "value currency" emerging from this integration. Simplistically put, you might imagine how each sense organ has its own way to tag information as potentially helpful or potentially harmful for viability. Feces or food trigger repulsion or attraction for the nose; a sharp rock or a lover's smooth body trigger repulsion or attraction for the skin; sight of a predator or a mate trigger repulsion or attraction for the eyes, etc. Integrating all these unique interpretive senses into a holistic unit, the self, with a single value currency—pain or pleasure—is necessary for the sort of conscious experience we are familiar with. When this happened as part of the evolutionary process, diverse meanings became unified into a single teleological, normative experience, and consciousness was born. Ginsburg and Jablonka write:

> There are constant changes in the world of a mobile animal, and it needs to detect them with its sense organs and respond to them through the flexible coordination of its muscle sheets. The value-reinforcement system has to accommodate these world and action changes and deal with their potentially conflicting effects. (Should the animal first satisfy its need for food and approach object X or retreat because it smells the scent of a predator in the area?) The animal must therefore have a "common (value) currency" for consistently evaluating *types* of world and body state over another type according to its value, which is context-dependent. As several scholars have suggested, pleasures and displeasures can be seen as such overall and general currencies of value, which can evaluate *any* percept or action.[76]

Consciousness, then, is the result of hierarchically integrating animal meanings in an enduring, temporally extended way. Experience is inherently evaluative, and inherently driven to detect meaning.[77]

Ginsburg and Jablonka compile a list of "neurobiologists' criteria" for consciousness that they say their account successfully meets. The list includes seven criteria which, I believe, can themselves be distilled down into just three categories: what I would call *intentionality*, *integration*, and *identity* (Table 1.2). Consciousness is the hierarchical integration of ententional learning processes into a single subjective identity experience with intentionality.

Table 1.2. Criteria for Consciousness

Intentionality	Attribution of values, emotions, goals; Selection, plasticity, learning, and attentional modulation
Integration	Global activity and differentiated states; Binding/fusion and unification
Identity	Embodiment and self; Temporal thickness

Consciousness, then, represents the holistic integration of ententional, meaning-driven cognitive mechanisms into a temporally extended manner, such that a distinct self emerges for whom such value states are *experienced* as subjective qualia like pleasure and pain.[78]

In this way, the meaningful information processed by cognitive entities as *good for* and *bad for* form the basis of nature's first subjective experiences, whose valences reflect the binary response patterns of what I prefer to call *attraction* and *repulsion*. Subjective valence states arise when this value-laden knowledge is networked in reference to the animal as a whole, to be broadcast coherently throughout the body in order to perform the requisite response as a unit that orients itself towards its goal states. Through this integration of information, the organism acquires the new functionality of the *experience of* "pleasure" (*good for*) and "pain" (*bad for*).[79]

Can the level of this integration be measured? The *integrated information theory* (IIT) of consciousness developed by Giulio Tononi and colleagues posits a specific metric, phi (Φ), for assessing how complexly wired or integrated the neurons of a nervous system are with one another, and directly identifies this measure with the level of sentient awareness of an animal. More conservatively, we can say that such integration levels likely *correlate* directly to degrees of

consciousness, in which case Φ level, if it could be directly measured, could help us quantify conscious depth.[80]

IIT posits a connection between an animal's neuronal complexity (Φ) and its general intelligence and problem-solving capacity, demonstrating the deep links between learning and consciousness.[81] Indeed, the increase in Φ across the evolutionary tree is helpful for mapping the complexification of animals on the plane of Mind, and can aid our survey of the complexification of meaning within the domain of Cognitive Learning. Christof Koch provides a visualization of this neuronal complexification across multiple species in his book *The Feeling of Life Itself* (Figure 1.10).

Figure 1.10. Increasing Consciousness and Intelligence in Animals[82]

The varying levels of neuronal integration in animals will thus translate into *different sorts of subjective experience*, and, with it,

different qualia systems that frame meaning in different ways relative to adaptive context. For instance, the mouse in Figure 1.10 will have a unique subjective experience that emerged through the integration of its species-specific sense for value states and goals relative to its environmental niche, typical food sources, etc. Simply (and clichély) put, cheese is meaningful for a mouse. The more complex dog will have its own (though evolutionarily related) values. A frisbee will be subjectively meaningful to it, but not to the mouse.[83] In short, different brains and nervous systems in different transjective contexts will be structured differently according to niche and need, though the underlying mechanics of meaning and relevance realization remain the same. All animals have their transjective values, their relative senses of meaning, which will be experienced differently at the subjective level and feel differently relative to neuronal complexity.

The issue of qualitatively distinct (and potentially incommunicably different) forms of consciousness for different beings was famously raised by philosopher Thomas Nagel in his oft-cited 1974 essay, "What Is It Like to Be a Bat?" In it, Nagel reflects on the ways that different animals' specific sense organs and ecological aims so color their subjective experience that it is difficult to see how one context-informed consciousness could even imagine the conditions of differently informed consciousnesses. "Our own experience," he writes,

> provides the basic material for our imagination, whose range is therefore limited. It will not help to try to imagine that one has webbing on one's arms, which enables one to fly around at dusk and dawn catching insects in one's

> mouth; that one has very poor vision, and perceives the surrounding world by a system of reflected high-frequency sound signals; and that one spends the day hanging upside down by one's feet in an attic. In so far as I can imagine this (which is not very far), it tells me only what it would be like for me to behave as a bat behaves. But that is not the question. I want to know what it is like for a bat to be a bat. Yet if I try to imagine this, I am restricted to the resources of my own mind, and those resources are inadequate to the task.[84]

Nagel is often cited as a seminal voice on this topic, but in fact such insights had already been anticipated a generation earlier by Jakob Johann von Uexküll,[85] a German biologist immensely influential on the fields of phenomenology, semiotics, and cybernetics.[86] Von Uexküll developed the idea that animals participate in a species-particular "*Umwelt*" (or "encompassing world") based on the specific kinds of sense organs they possess and the capacities for action and cognition these afford. To use Nagel's example, a bat perceives the world primarily through sound, not sight, by means of which it is able to locate its food. For "bat consciousness," then, information is processed in a manner quiet alien to us, with feelings and normative goals quite foreign to those of humans. Based on its neuronal makeup, which evolved according to meet the needs of its environmental niche, bats have learned their own forms of meaning. Indeed, as von Uexküll insightfully argued, *the Umwelt of an animal is defined by what matters to it.* An animal interprets its meaning based on what it can perceive and do, and what it can perceive and do delimits its range of meaning:

> Each environment [Umwelt] forms a self-enclosed unit, which is governed in all its parts by its meaning for the subject. According to its meaning for the animal, the life stage includes greater or smaller space, in which the places are completely dependent in number and size upon the capacity of the sense organs of respective subjects to draw distinctions. …Anything and everything that comes under the spell of an environment [Umwelt] is either redirected and re-formed until it becomes a useful carrier of meaning or it is completely neglected.[87]

Such is von Uexküll's "theory of meaning," which arises out of his biologically-informed phenomenological considerations of different animal consciousnesses.[88]

From all this, we can conclude that meaning and value are circumscribed by the consciousness of the animal as it has been shaped through adaptation to its environmental conditions. Meanings and values emerge through the transjective system–environment relationship, with an objective aspect that goes all the way down to thermodynamic necessity and a subjective aspect colored by the values constitutive of the exchange.[89]

In some senses, von Uexküll was himself anticipated in this recognition by an even earlier thinker, anthropologist and philosopher Ludwig Feuerbach. Feuerbach intimated the importance of such dynamics when, as early as 1841, he wrote:

> If the plants had eyes, taste, and judgment, each plant would declare its own flower the most beautiful; for its comprehension, its taste, would reach no farther than its

> natural power of production. What the productive power of its nature has brought forth as the highest, that must also its taste, its judgment, recognise and affirm as the highest.[90]

Feuerbach makes this case as much for the caterpillar as the human being. The limits of the nature and consciousness of a being define the potential limits of the value and meaning it can cognize.

When we combine these older insights with those of contemporary neuroscience, we can actually begin to link these qualitative distinctions with quantifiable data. Having seen that the level of neuronal complexity plays a crucial role in the origin and evolution of consciousness, we can relate the growth in brain size and integration to the complexification of meaning over time. More integrated nervous systems lead to more complex minds, whose Umwelts will be more expansive in sensory input and active capacity. *The evolution of meaning parallels the evolution of consciousness.*

One clear example of this appears in the emergence of social values among animals. With the deepening of subjective, felt experience, reproduction and the rearing of offspring become richly *emotional* activities. More neuronally complex, intelligent mammals form social bonds and relationships with other members of their group that are mediated by value-laden informational exchanges. That is, relationships with other animals have considerable *meaning* to them, and this is mediated via strong qualia states. One obvious instance of this is to be found in the mother-child bond of advanced primates, but dolphins, elephants, and whales have also been documented

exhibiting strong social emotional bonding relationships.[91] Such emotional investment into social dynamics also plays out in dominance hierarchies, as males strive to assume positions in the social hierarchy that will render their genetic continuance more viable (e.g., becoming the alpha in a group of gorillas).

This extension of value and meaning into the social domain opens up new levels of complexity that animals must navigate. As we have seen, complexity begets more complexity, as more complex environmental conditions require more complex adaptive solutions. So, as complexifying nervous systems meant bigger brains that could navigate social environments, those new environmental conditions fueled a drive towards even more neuronal complexity. In such cases, evolution selected for bigger brains with even more integrated neuronal activity and conscious depth.

This "encephalization" process of generating bigger and bigger brains, then, can be read as an empirical indicator of meaning's complexification—the spatio-structural flipside of rising Φ and thus deepening subjective consciousness, with increasing intelligence orienting its growing learning capacities towards a more expansive field of value. We see it characterizing the evolution of social primates up to advanced hominids, when the process suddenly explodes exponentially around *Homo habilis* (Figure 1.11).

Why this sudden takeoff? Had some new variable emerged from this trend, catalyzing it further? Some novel phase transition to a still higher plane of complexity?

Could this inflection point signal the dawn of *language*?

Figure 1.11. Increasing Brain Size of Hominids[92]

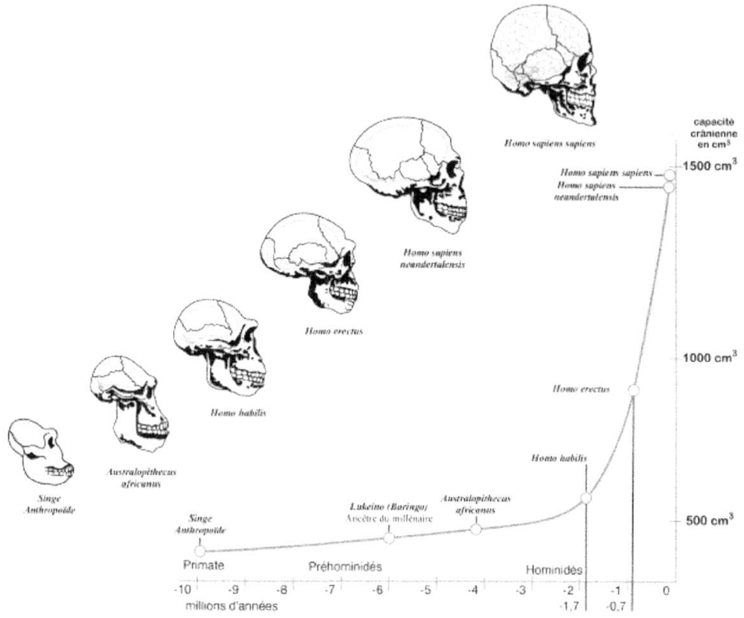

In function, symbolic language appears as a natural extension of earlier Cognitive Learning processes attentive to meaning. Henriques notes:

> [T]he human mind is a neurocognitive behavioral investment system that symbolically tags objects and changes in the form of nouns and verbs. …[T]the neurocognitive architecture of the preverbal human mind…is structured to: (a) attend to relevant objects in the environment; (b) identify changes that are happening in predictable and unpredictable ways; (c) determine what are desirable and undesirable outcomes; and (d) direct subsequent work effort (i.e., behavioral investment) in an

> attempt to realize desirable outcomes and avoid undesired ones based on probabilistic expectations. In short, evolution built our primate phenomenology to see the world in behavioral terms. In addition, nature added a language acquisition device…that allowed us humans to generate a symbolic-syntactical representation of our phenomenological experience.[93]

In this way, language emerges directly out of our pre-existing Cognitive Learning capacities to mentally represent the world; which, we have seen, emerged out of our Genetic Learning capacities to biologically encode knowledge of the world in our genome; which emerged out of Structural Learning capacities to maintain informational identity by adapting to the energic pressures of the environment. Symbolic communication simply continues the universal learning process into a new register through the introduction of a novel information processing system: *language.*

As Deacon notes in his 1997 book *The Symbolic Species: The Co-evolution of Language and the Brain*, the emergence of an early proto-language seems to have thrown fuel on the fire of the encephalization process, further complexifying the mind. He writes:

> The remarkable expansion of the brain that took place in human evolution, and indirectly produced prefrontal expansion was not the cause of symbolic language but a consequence of it. As experiments with chimpanzees demonstrate, under optimal training conditions they are capable of learning to use a simple symbol system. So, it

> is not inconceivable that the first step across the symbolic threshold was made by an australopithecine with roughly the cognitive capabilities of a modern chimpanzee, and that this initiated a complicated history of back-and-forth escalations in which symbol use selected for greater prefrontalization, more efficient articulatory and auditory capacities, and probably a suite of other ancillary capacities and predispositions which eased the acquisition and use of this new tool of communication and thought. Each assimilated change enabled even more complex symbol systems to be acquired and used, and in turn selected for greater prefrontalization, and so on.[94]

We need not get mired down here in considerations of just how and why language evolved—a question that is still much debated and beyond the scope of this book series. I would simply reiterate the point, noncontroversial among scholars, that it appears to have been directly linked to increasing neuronal complexification (i.e., bigger, more densely networked brains), such that Symbolic Learning can be seen as a byproduct of Cognitive Learning. It is also worth noting that some theorists, such as evolutionary biologist Kevin Laland, see language's origins specifically in *social learning dynamics*.[95] Such a theory fits nicely with the picture I have been painting of a continuous universal learning process leading to the generation of successive, novel information processing systems out of earlier ones.

Whatever its origin, though, it is clear that, once begun, symbolic language precipitated an autocatalytic process leading to a profound qualitative shift in the lived reality of *Homo*

sapiens. For out of this transformation came the truly essential aspects of the human experience: namely, *the self-conscious individual ego that constitutes the human person as well as the conceptually-mediated technologies, institutions, and collective visions that constitute collective human culture*. With these, a whole new kind of meaning emerged in the world—the kind that will preoccupy us for the remainder of this book series.

4. Symbolic Learning (Culture)

Finally, we come to the sort of semantic information processing unique to human beings: Symbolic Learning. Such learning is distinct from all other kinds in that it participates in inter-subjective socio-linguistic contexts. The day a child is born, it becomes immersed in a linguistic world, a cultural Umwelt of shared *words*, *concepts*, *ideas*, etc. Indeed, long before it can actively participate in this discursive world itself, a child is spoken to, sung to, and made a passive recipient of conversations, symbols, social customs, and so forth. By means of these, a person will gradually learn to navigate their world through reference to shared abstract symbols.

Such symbolic information is hardly the invention of just the rearing parents, of course, but predates and transcends them, too. Child-rearers are largely passing on what they themselves learned in their own rearing and cultural immersion. Education, from the beginning, aims to initiate humans into a transpersonal socio-linguistic world. This is called *enculturation* or *socialization*: the process by which human

offspring are slowly molded into active participants within their linguistically-mediated culture.

The period of educational dependency lasts far longer for human children than for other animals in large part because of the specific demands of such a learning process.[96] During this time, children have to learn not only the kinds of sensory-motor skills that other animals develop to physically navigate the world, *as well as* the group dynamics of complex social mammals, but *also* the intricacies of the purely symbolic world of their culture's meaningful concepts, ideas, stories, norms, etc.

The symbolic language that enables such mental activity moves beyond mere ostensive reference (e.g., communicative signals tagged to specific phenomenal signs, such as animals use) to an entirely new register of cognitive organization. This is marked by the appearance of *grammar* and true *propositional* claims. Not just "gazelle, there," but "a gazelle *is* there."[97] This important transition, argues Henriques, marks the advent of Culture proper, as it initiates the uniquely human dynamics of *justification.*

According to Justification Systems Theory (the UTOK metatheoretical framework for understanding the basis of the Culture plane of complexity; see Figure 1.7), human language's development of propositional claims opened up a new discursive space of claim and question, argument and counter-argument. With the linguistic declaration "a gazelle is there" comes potential rejoinders: "Is it?" "Are you sure?" "How do you know?" etc. This development from animal communication to propositional justificatory language set off a cascade of adaptive pressures and opportunities that would

radically transform the Umwelt of *Homo sapiens*, marking the transition from minded animals to genuinely enculturated *persons*.

Justification Systems Theory is an ambitious metatheory that would account for those aspects that make human beings unique from other animals by revealing how the emergence of language utterly reshaped human consciousness and meaning-making. Its three core claims Henriques summarizes this way:

> (1) the evolution of propositional language generated question–answer dynamics framed as the adaptive problem of justification, and this in turn shaped both the evolution of human self-consciousness and human culture such that; (2) the nature of human consciousness can be framed by the "Updated Tripartite Model" that differentiates it into three different domains of (a) the experiential self, (b) the private ego, and (c) the public persona; and (3) humans are transformed into persons as a function of being socialized into systems of justification, such that, relative to other animals, human persons operate in a new complex adaptive plane that can effectively be framed as Culture with a capital "C."[98]

Understanding these developments is key if we are to appreciate how the meaningful information processed through Symbolic Learning emerges out of and builds upon that of Cognitive Learning. So let's unpack them—paying special attention, as we do, to the meaning dynamics as they unfold.

(1) Animals, we have seen, are goal-directed, aim-driven entities. Their complex nervous systems allow them to develop

and store mental models of reality, physically navigate through those simulated environments, and provide the conscious animal as a whole with ongoing emotional feedback on their actions to help direct its ongoing decision-making and energetic investment. For social animals whose survival depends on successfully operating in collective groups, these Cognitive Learning processes are naturally extended to include mappings of social dynamics as well. Meaning becomes more "indirect" as survival is here contingent on attending not just to one's immediate needs and goals but to the *collective's* needs and goals as an extension and support of one's own flourishing. Meaningful information now includes how to optimally navigate the behavior of other animals with *their* goals and intentions.

Without language, though, even social animals remain completely isolated within their own individual meaning-making frame—the subjective conscious experience of an animal's normatively-directed emotional life. The intentions of others remain opaque, gleaned only through outwardly manifested behaviors (e.g., a powerful blow, a mating dance, etc.). Though mapping a social world, meaning is processed only by and for individuals through the neuronal information processing system of those individual animals.

But symbolic language changes this entirely. How? By tearing down the wall isolating meaning within the mind of the individual and opening up the world of meaning to *inter*-subjective exchange. Not just the outward *behavior* but the originary, motivating, subjective *meaning* content of an individual mind can now become a shared object of concern. Subjective understanding and intention become interpersonal

understanding and intention as the inner world of conscious reasons breaks its confines with a single word: "Why?"

With such an advance, a cosmos of brute, declarative fact learned the *interrogative*—and meaning slipped the prison of the solipsistic self. Henriques writes:

> Unlike chimps, language allows humans to ask and be asked about the thought processes associated with their behaviors. Questions such as "Why did you do that?; What gives you the right to behave that way?; Why should I trust you?" force the issue. Obtaining information about what others think, what they have done, what they plan to do is obviously important for navigating the social environment in modern times and given that humans have always been an intensely social species, there is every reason to believe that it was equally essential in the ancestral past. So once people developed the capacity to use language to access the thought patterns of others, they likely did so with vigor.[99]

The question–answer dynamic this birthed is entirely novel to the Culture plane of complexity, rendered possible by the unique nature of symbolic information processing, which is able to "move between" thinking agents in a way completely impossible for mere neuronal processing. Whereas Genetic Learning occurs at the phylogenetic level of the collective species, and Cognitive Learning at the ontogenetic level of the individual organism, *Symbolic Learning allows ontogenetic knowledge to move between individual and collective levels*. As Henriques puts it:

> Language-based thoughts, however, are different. This is because propositions move through the nervous system and the boundary of the skin and out into the world and back again without losing their informational form. ...Our capacity to share in language means that a direct highway of mental information is opened between us. ...*Although I cannot directly share with you my subjective conscious experience of redness, I can directly share my self-conscious, language-based description of redness and any other language-based thoughts I have.*[100]

Simply put, symbolic language renders inter-subjective communication possible. It allows me to translate the cognitive information processing I experience internally into externally transmittable data that others can incorporate into their own learning. Justification is *meaning-sharing.* It is made possible by the ability to transfer meanings processed through Cognitive Learning into an externalizable symbolic language that encodes our reasons and intentions in a manner others can decipher and engage. In the process, a novel information processing system emerges and meaning becomes collectively linguistic.

(2) Such meaning-sharing, though, places a new burden on the individual, for it demands that they first be sufficiently *self-conscious* of their own internal cognitive processing and personal meanings if they are to properly translate them into language for others. That is, I must develop a self-reflective stance towards my own internal cognitive experience and decision-making—a higher-order perspective on my own intentions, wants, desires, etc.—if I am to model them to others using language. With this demand, a new kind of

individual had to develop: the *self-conscious person.* If consciousness emerged when animal value systems became integrated under a higher-order organization of a singular "I" that could recursively relate to itself through self-modeling, then *self-consciousness* emerges when that conscious totality loops back yet again on itself to model itself through language.[101] So out of the unreflective experiential self is born the "ego."

Henriques's model of the human psyche emerges from this crucial insight. Each person has a pre-linguistic subjective experience like other animals who operate on the plane of Mind. But language affords us the means of rendering this experience communicable, both to ourselves and to others, in pursuance of our normative goals. Doing so leads to the production of a personal ego, which is that private, internal "voice" continually narrating and justifying our experience to ourselves in linguistic terms.

Such egos are inherently linguistic in nature—an internal narrative of self-justification continually translating raw emotional experience into the language of symbolic speech. I do not just feel hungry and eat. I think, *"I'm hungry. I'm going to get something to eat."* More than that, I engage in an ongoing internal monologue of justifications in such terms: *"Should I eat the last slice of pizza in the fridge? I think my brother was saving it, so I probably shouldn't. Then again, maybe it's not that important to him? Plus, he wouldn't know it was me… No, I'll leave it for him. He seemed interested in it. Besides, he'd definitely assume I did it and get mad at me and then it'd be a whole thing. I'll just make a sandwich."* In pursuing my own goals, it is as though I have an internalized other person continually demanding justification for my actions—an enculturated voice in my head asking how my reasons and

meanings mesh with others'.[102] The stories I tell myself about myself to organize and justify my behavior, *that's* my ego.

For many reasons, however, individuals do not communicate to others the full contents of their egoic thoughts. Individuals' goals, including maximizing social influence and prestige, are often better pursued by selectively filtering publicly communicated thoughts from privately held ones. The strategic version one presents of themselves to the world, then, is their *persona.*

The resulting three-part conception of the human individual is thus composed of (a) an *experiential self* (the conscious animal mind), (b) an *ego* (a linguistically-mediated narrative reflection on the experiential self), and (c) a *persona* (the self-concept one projects to others in an effort to maximize success in their social context). Henriques calls this an "update" to the older "tripartite self" (cf. Freud's id, ego, and superego), and one that does justice to insights from both folk and academic psychology by placing all within a metatheoretical framework rooted in the planes of complexity that have emerged over cosmic evolution (Figure 1.12).

Figure 1.12. Dynamics of the Human Psyche in Culture[103]

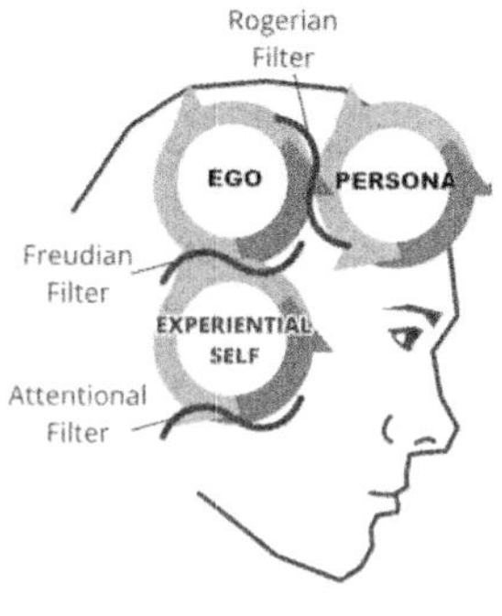

(3) The emergence of language as a means of justification, then, profoundly altered the Umwelt of *Homo sapiens* as the new dynamics transformed the entity–field relationship from "animal–environment" to "person–society" by linking self-conscious, self-justifying egos to other self-conscious, self-justifying egos through meaning-sharing. Just as animal Mind represented a networking of meaning-driven cellular Life to produce an emergent plane of complexity, so human Culture represents a networking of meaning-driven *minds* to produce an emergent plane of complexity. Culture, then, is the product of many conscious animals using symbolic communication to integrate subjective information into an inter-subjective world model with *collective normative goal-states.*

Indeed, this networking of shared meanings is why we must speak of justification *systems.* As an inherently cultural, interpersonal phenomenon, justification never occurs in isolation. Just as there are no "private languages" (as Wittgenstein observed), there are no private justifications.[104] Linguistic meaning-sharing inherently cuts across levels and scales, with person-to-person transjective relationships embedded in micro-, meso-, and macro-level justification contexts.

Drawing on Urie Bronfenbrenner's Human Ecology Theory,[105] Henriques frames this as a nesting of more immediate, local contexts within broader, more encompassing ones, with every one bearing on the other (Figure 1.13).

Figure 1.13. Bronfenbrenner's Contexts of Development[106]

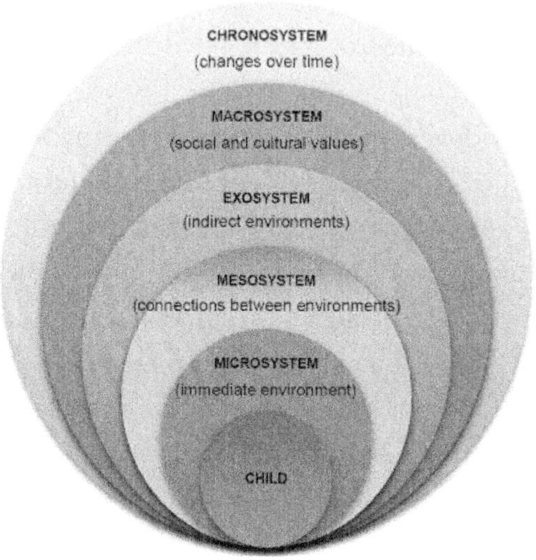

When we consider the sort of mind-networking that is implicit in the socialization of human egos within linguistically-mediated Culture, then, we are necessarily considering a collective, transpersonal context for human meaning-formation that transcends individuals. It is the individual-sustaining *group* that is very much the meaning-defining entity now, similar to the way the animal becomes the meaning-defining entity when cells network together into a multicellular organism. The cell-level meaning does not vanish; but it does shift to accommodate a broader context and serve the normative needs of a broader whole. With the emergence of Culture, meaning becomes networked into a macro-level justification system that serves to orient the goals of the collective and keep *it* far from equilibrium. *Meaning at this highest-level macro-scale is called a worldview.*

Worldviews are collective justification systems that serve to direct the normative goals of society as a whole. From this vantage, Azarian writes,

> An individual's value is not about physical ability or intelligence; it is about how one contributes to preserving the stability of the social organism we call society. By being unique, or having a unique perspective that you share with others through conversation, writing, or art, you are contributing to the diversity of ideas that collectively create the world model of the superorganism. For social organisms, we call that world model a *cultural worldview*—which can be a religion, national identity, political ideology, or cosmic philosophy. We can make use of metaphor by saying that *the worldview is the Markov blanket for the social organism*, because it is what binds the independent agents together to form a collective whole with internal states that are distinct from the external states of the environment, even though there is no tangible physical barrier, like a cell membrane or an integumentary system.[107]

So, just as the individual cells of the nervous system all contribute to the configuration of a world model that aids the viability of an animal in context, so individual human minds in a collective justification system all contribute normative inputs to the world model of the group. "What is the biological function of a cultural worldview?" asks Azarian. "The worldview, like the world models that came before it, is a

strategy for survival, but at this scale it is a strategy for *collective survival*, for the continued persistence of the superorganism."[108]

With this framework in place, we can begin to appreciate the dynamics of Symbolic Learning as a specific manifestation of the universal learning process. For, as Azarian suggests, the value of individuals is their ability to contribute diverse and novel inputs to the collective world model. When they do, they *update* the model the same way the nervous system updates its model in the face of novelty and surprise. Collectives can *learn*, adjusting linguistic-conceptual systems of understanding and justification by integrating new information. This is how worldviews evolve, transforming normative frameworks of meaning to adapt to new circumstances and information. And since the overall direction of evolution is towards more optimal flows of energy and informational complexity, worldviews of Culture tend to complexify just as Life and Mind complexify.

That is the macro-scale picture of Symbolic Learning—the complexification of cultural meanings, which impinge upon and influence justificatory contexts below it, all the way down to the micro-scale of the individual, whose ego takes shape *through enculturation into that shared worldview of the group*. At this level, Symbolic Learning is an ontogenetic process, unfolding through individual adaptive interactions with the environment. Here, despite language's inherently collectivizing nature, meaning is still individual, concerned with the viability of the person in context. As the "unit" of Culture, the ego learns to meaningfully navigate its world to achieve its normative goals as a multilayered entity very, very far from equilibrium.

The focus of the remainder of this book series will be the transformation of human meaning systems through Symbolic Learning. Given the inherently transpersonal nature of the linguistic medium through which it operates, this will require looking at both micro and macro scales—that is, the ontogenetic learning done by individuals and the phylogenetic learning done by the species through cultural evolution. Through ontogenetic learning, individuals generate their own, slightly different world models. Through justification dynamics, individual decision-making and reasoning based on these world models become subject to public scrutiny and winnowed through the selection mechanisms of Symbolic Learning. In this way, human collectives learn to "think together," cohering collective justification systems through the integration of mental information processed by its members. Individuals feed symbolic information to the collective and the collective rears individuals by means of established symbolic information. *That is what enculturation is*. Through this process, mediated through individual learning, worldviews develop and human meaning evolves.[109]

These dynamics will be explored in much greater detail when we consider *the individual/collective feedback cycle* in volume three. For now, we conclude by noting that knowledge creation in the cultural world is an extension of the integrative evolutionary synthesis that describes cumulative learning at *pre-cultural* levels of reality.[110] When we follow that trajectory into the plane of Culture, we see that human intellectual history is the linguistic version of a knowledge generation mechanism driving the entire complexification of the cosmos.

Meaning across Scales

So, what can we say about meaning's "complexification" then? Throughout this book, I have endeavored to show how meaning has indeed grown richer and more informationally rich across cosmic evolution. Though meaning ultimately stems from the existential requirement to resist entropic dissolution, emergent meanings of complex entities clearly cannot be simplistically *reduced* to this foundation but recognized as qualitatively transforming as complexity increases. As meaning has complexified from Matter to Life to Mind, new layers added upon existing ones, building upon previous developments. New kinds of entity–field relationships formed, with new kinds of information to process and interpret for meaning. In this way, genetic meanings gave rise to neuronal meanings, which (for one group of primates with exceptional neuronal complexity) eventually gave rise to *language* and the domain of symbolic meaning: the stuff of Culture.

Along the way, meaningful concerns and behaviors proliferated. Beginning as a purely thermodynamic imperative to materially persist, meaning deepened into complex organisms' goal-directed behavior of energy capture, self-maintenance, and reproduction; then infinitely diversified through an explosion of mobile animals normatively navigating a vast possibility space by means of increasingly complex mental models. All this eventually gave rise to (and plays out within) an *inner* world of subjective experience, wherein normative values became *felt* as qualia states for a persisting experiential self—which in time led to social selves

in group dynamics sensitive to complex social relationships of all sorts.

Language both reframed meanings from previous levels and opened up entirely new domains of meaningful experience. For instance, it *aided* humans' continued material existence by enriching our mental models with abstract concepts, whose patterns dramatically increase our competencies for action and thus our viability. It *enhanced* and *deepened* social relationships through intersubjective communication, collective problem-solving, and shared group goals, all of which reshaped sexual relationships and power hierarchies. In these and a million other ways, language transfigured meanings from the earlier levels of Matter, Life, and Mind. But it also opens up entirely *new* realms of meaning unique to the level of Culture, as language gives rise to a concept of an egoic self, a theory of mind, and entirely abstract concepts such that meaning can now unfold through self-awareness, relationship to other selves, and the vast horizon of human reflection, from philosophy to logic to religion and science and beyond. With language, the full range of human meaning becomes possible.

Jumping back to our primary sense of meaning discussed at the outset—that of semantic information as what aids an entity in resisting entropy—we see that a throughline does indeed run from it all the way to the rich and diverse activities of cultural life that we ourselves actually tend to identify as meaningful and valuable in the common colloquial sense. However, to be clear once more (in case there is any lingering confusion), this is not at all to suggest that human meaning can be simply *reduced* to a survival imperative, any more than human

anatomy can be reduced to the skeleton or the contents of a book can be reduced to the paper it is written on. Rather, this throughline simply shows how, without the skeleton, the body cannot stand; without the paper, the book could not be read.

From a secure, naturalistic foundation of meaning in thermodynamic principles, we have thus been able to build up successively more complex layers spanning multiple novel information processing systems which are dependent upon (but not reducible to) the substrates from which they arise. By the time we reach the Culture level of human meaning, there is a rich tapestry of various semantic registers interweaving symbolic, social, experiential, genetic, and other sources of meaningful information. Underpinning them all, however, is the same basic teleological and thus normative endeavor of successfully persisting and enhancing viability. As physicist Carlo Rovelli puts it in a 2018 paper, this most basic notion of meaning is the first link in a long chain of successively more complex meanings. "The claim here," he says,

> is not that the full content of what we call intentionality, meaning, purpose—say in human psychology, or linguistics—is nothing else than the meaningful information defined here [as thermodynamic viability]. But it is that these notions can be built upon the notion of meaningful information step by step, adding the articulation proper to our neural, mental, linguistic, social, etcetera, complexity. In other words, I am not claiming to provide here the full chain from physics to mental, but rather the crucial first link of the chain.[111]

Rovelli distinguishes between "direct" and "indirect" meaning to reflect the difference of semantic information that has immediate causal bearing on viability in a thermodynamic sense and that which only secondarily relates to it (e.g., by being relevant for obtaining more existentially relevant information). "Obviously," he writes,

> not any manifestation of meaning, purpose, intentionality or value is directly meaningful… Reading today's news paper is not likely to directly enhance mine or my genes' survival probability. This is the sense of the distinction between 'direct' meaningful information and meaningful information. The second includes all relative information which in turn increases the probability of acquiring meaningful information. This opens the door to recursive growth of meaningful information and arbitrary increase of semantic complexity. It is this secondary recursive growth that grounds the use of meaningful information in the brain. Starting with meaningful information in the sense defined here [as what aids an entity in resisting entropy], we get something that looks more and more like the full notions of meaning we use in various contexts, by adding articulations and moving up to contexts where there is a brain, language, society, norms…[112]

Over the course of this book, I have endeavored to do just this: move up the various levels coupling entities to environmental contexts in more complex relationships, whereby more complex meaning is generated. By the time we reach the domain of Culture and human symbolic meaning, we

are in an intricate field indeed, where all sorts of information vie for attention and interpretation, both directly and indirectly.

What all of this ultimately offers us is a robust theorization of human meaning and value, and a framework for understanding how and why these emerge and change. It allows us to see how 1) meaning in human culture constitutes a dynamic system of semantic information processing at multiple levels, linking the human person to material, biological, neuronal, and cultural realities; 2) these meanings play out *in context*, in response to specific conditions that characterize the entity–field relationship; 3) these meanings are subject to the continual revision of an ongoing learning process, whereby new information reduces uncertainty and reshapes beliefs and values.

In an attempt to represent this cumulative and dynamically recontextualizing nature of meaning across cosmic complexification, Table 1.3 presents the four types of learning in relation to the new layers of meaning that accrete with their novel forms of information processing.

Table 1.3. The Complexification of Meaning

Complexity	Learning Type	Information Processing Medium	Meaning
Less	Structural	(Wave function)[113]	*Persist*
↓	Genetic	DNA	*Maintain, repair, reproduce*
↓	Cognitive	Neurons	*Model the world accurately* *Feel good* *Foster social relationships*
More	Symbolic	Words	*Relate inter-subjectively* *Deepen self-awareness* *Transcend conceptual limitations*

Such meanings, we see, constitute an unbroken chain that leads from the most rudimentary to the most lofty.[114] At every level, meaning depends upon the various lower-order meanings that support it. Fostering social relationships, for instance, is rooted in the valence qualia of consciousness, which emerges from world modelling, which depends upon being alive, etc. The meanings unique to the human Culture plane given here are, however, admittedly little more than placeholders for the discussion that will follow. It will be the aim of the remainder of this book series to spell out more fully just what the evolution of human meaning and value actually looks like as Culture complexifies.

To get a better appreciation for how meaning can complexify in this way, let us look at a couple of examples. For this, we can use two profoundly meaningful issues operant at all scales of complexity: namely, *generation* and *annihilation*. Such matters are, obviously, fundamental to the issue of viability. What I wish to show here is just how these meanings evolve through complexifying Umwelts.

Generation at the level of Matter is basic and blunt. Nuclear force attractions draw protons and neutrons together; gravity clumps hydrogen atoms; atoms amass as molecules, etc. Generation is brute and binary; it is, or it is not. It happens, or it doesn't—the product of chance or determinism. Energy gradients push matter to self-organize as conditions allow, but structures dissolve as conditions change.

At the level of Life, generation becomes *reproduction*. Cells divide and duplicate—a marvel of organization that lets Life's information live on. Indeed, that is the functional purpose of biological generation. Then, sometime around two billion years

ago, sexual reproduction evolved, wherein two different living organisms come together to combine their DNA to create a third. On the complexity plane of Life, sex is purely mechanical; the "knowledge" to sexually reproduce exists only within the genetic code of the different sexed organisms. In flowering plants, for instance, the stamen is the male part that produces seed and the pistil the female part with the ovary. Seed makes its way to the ovary by wind, pollinators, or other means. Sex is an unconscious and unexperienced act, as there is no nervous system to mediate consciousness or experience.

This all changes at the level of Mind, where sex becomes a behavior of animals. At this level, the meaning of sex becomes infinitely richer, as it is now overlaid with intense sensory valence qualia thanks to complex nervous systems. Generation becomes *mating*, operating in a whole new domain of behavioral rituals, performance, competition, etc. Sex becomes carnal and embodied, mediated through pheromones and desires. At less complex levels, the brute, mechanical nature remains dominant. Some insects devour their mates, while some animals accidently kill theirs in the forceful act of reproduction or competition for it. The drive to reproduce is intense, compulsory, heedless. In more complex animals like mammals, mating is woven into the social dynamics of the group. It becomes attached to status (as in chimps), or the act is offered outside of reproductive purposes purely to serve social relationships (as in bonobos).

At the level of Culture, sex deepens even further in its manifestation among self-conscious persons. Symbolic communication opens inter-subjective channels for a novel kind of intimacy between egoic individuals unseen in the

animal world. The complexities of language allow the idiosyncratic nature of personality to become a factor in selection. Brains and not just brawn become important. In colloquial terms, animal lust is overlaid with human love, such that a profound reciprocity of mutual care and satisfaction can now exist. Sex attains a "spiritual" depth, but is also refracted through cultural taboos and societal norms. The typical physical sex distinctions of animals (male/female) get an overlay of cultural gender constructs (masculine/feminine), performed or subverted differently across contexts. Sex can be play, punishment, heaven or hell, rape or rapture.

In short, the meaning of sex to a human in the 21st century is quite different compared to that of, say, a dandelion, despite the continuity of the mechanism. Indeed, not only are there the successive layerings of neuronal and symbolic information from the Mind and Culture levels, respectively, there are also *intra*-Cultural layerings, as human Symbolic Learning has shifted cultural perceptions of sexuality over the millennia. Sex means something different to a postmodern woman than to a woman in a Neolithic forager society, a woman in a traditional agricultural society, etc. *That* it is meaningful has not changed, only *how* it is meaningful has evolved.

On the other end of the spectrum, *annihilation* can be assessed in similar terms. For Matter, annihilation is simply the binary 0 to generation's 1. Stars explode with all the mute indifference with which they formed. Entropy is simply the way of things.

For cellular Life, waging its war against entropy, annihilation means *death*, its antithetical goal state. For cells, the avoidance of death is a reflex, an unconscious reaction

triggered by a stimulus. The amoeba tracks to the left when it bumps up against a destructive chemical on its right. When dissolution does come, the specific system is lost, but near-identical copies exist elsewhere. So Life flourishes unconsciously, and disappears into darkness just as unawares.

For animals, death is profoundly different. Confronted by danger, there are feelings of fear, stress, terror. An animal will struggle violently to preserve itself. All instincts marshal to survive. When death is finally imminent, it is *painful.* From the intimation to the wound to the final undoing, all qualia are negative; the animal *suffers.* Once dead, the unique aspect of that being, the product of all its experience, is lost to the world. Some social animals, like elephants and apes, seem to recognize this, and appear to mourn, in their way, the unique character that has been lost.

On the plane of Culture, all of this is only heightened by symbolic thought and action. Death is mediated by ritual, by grief, by concepts of afterlife and ancestor worship. Death can *mean* a great deal, depending on the cultural overlay of interpretation. An individual not only experiences death but can also *reflect* upon it. Indeed, the prospect of death can terrify a human entirely in the abstract, when no specific danger is imminent at all. One can wallow in the dread and uncertainty of it, and even consider inflicting it on oneself. One can ask, "To be or not to be?" One can be oppressed by the weight of the realization that all that is living must die, and persist in the torment of the ever grinning skull latent in every pleasure. A person can also heroically face this inevitability with courage or acts of self-sacrifice. Aware of suffering, one can seek its mitigation in the world. One can apply one's thought to reduce

death and negative valence. People can also rationally accept death, come to peace with it as no other animal can. They can assess the sum total of their experience and conclude, amidst the pain of dissolution, that it was all "worth it." Humans can thus go gentle or not into that good night, forking lightning as best they can with their allotted time. They can imagine heavens and hades of every stripe, which ease their passing and gesture to a viability beyond the physical. But humans do pass, and the patterned information of the infinite richness of their egos is experienced no more by those they leave behind. Death is thus bitterest for the most complex beings, who have the depth of consciousness to taste the nuanced flavors of its bitterness. Death is "the last enemy,"[115] but also "the mother of beauty."[116] A complex issue indeed.

Thus, from fundamental thermodynamic laws we see the blossoming of ever richer phenomena through novel information processing. This is how meaning evolves. In that evolution, we can find both continuity and discontinuity. New layers are not reducible to lower ones but do build off of them. There is a thread connecting the chaotic dissolution of a whirlpool and Hamlet reflecting on "the undiscovered country"; it is a tenuous thread, but it meanders the labyrinth of complexity as surely as Ariadne's. In the symphonic opera of existence, meanings are the leitmotifs, appearing soft and solo at the first, then rising, growing, getting contrapuntal harmonies and spreading through the orchestra, until the single melody that broke the silence thunders with a swell of ecstasy, sublimity, and awe all past our words.

The theme of the symphony is what we call *the sacred.*

The Sacred

According to the concept of meaning and value advanced over the preceding pages, complex entities strive to update their world models through learning, which provides them the relevant information they need to orient their actions towards optimal goal states. As they do, meaningful knowledge is attained that permits the achievement of still greater levels of complexity, and so on. This process plays out in human culture through the development of symbolic meaning-sharing, which is a form of semantic information processing able to intersubjectively cohere large groups of individuals in a symbolic world dedicated to achieving collective goals.

As meaningful knowledge is learned by people in societies, it is symbolically translated and culturally affirmed, whereupon it serves to *in-form* future action and collective organization. A "sense *for* meaning" is thus cultivated among a society's members, a honing of predictive accuracy in the direction established by earlier successes, as new information is continually compared against the proven templates of a society's established meanings. Language allows these established meanings to be reified and reflected upon, and through collective cultural memory these meanings are stored and transmitted to new generations across time via the enculturation process discussed above.

Such meanings will naturally vary in their cultural significance relative to how important they have been for steering society towards continued or enhanced viability. Novel and unfamiliar concepts and behaviors will face, if not ambivalence, perhaps only mild curiosity. More established

lodestars will receive varying degrees of regard. But for the learned meanings established as absolutely fundamental to orienting a society towards locally optimal states, there will be especial reverence, care, and devotion. Such meanings we call *sacred.* And it is according to their "gold standard" that all other collective meanings and justifications are weighed and assessed.[117]

"The sacred," then, can be understood as both the symbolic repository of a culture's most proven meanings and values *as well as* the informing template or ideal by means of which people assess novel information for meaning. Cultural memory serves, you could say, as a transpersonally aggregated world model by means of which enculturated individuals learn to predict meaning. Embodied predictive processing of cultural meaning with the greatest causal power manifests as intuitions of sacredness.[118]

Over time, such collective learning carves out a hierarchy of values relative to the specific environmental conditions of the society. Like the formation of a neural network, the most successful meanings and values become well-trodden and robust, and get deposited into the symbolic social repository as sacred beliefs. The less established, exploratory pathways remain weak, and could as easily lead to failure as to new, more successful beliefs. Symbolic Learning is the process by which cultures update their collective world models towards more viable beliefs, in the course of which the sacred itself is updated. Because the sacred is both what is updated as well as the guide to more robust meaning, there is both disruption and continuity in this learning process. The sacred does not just

change, it *develops.* This book series is an attempt to chart that development.

I will use the word "sacred," then, to refer to those collective representations of a culture that are superlatively meaningful to it (in the technical sense of meaning as mutual information with causal power bearing on viability). The sacred constitutes the veritable foundation upon which psycho-cultural flourishing depends. It hallows that which sustains and upholds a given society, enshrining the established gains of cognition and conception that make any continued gains possible. It is the proper respect owed to the adaptive path taken, its active shoring up, and the intimation of where it will lead based on where it has been. It is, in short, what is deemed most important to the continued viability of a successful human system–environment relationship.[119]

One glimpses this sense even in colloquial uses of the word, as the "sacred" has been extended beyond traditional religious discourse into other domains of cultural significance. A cathedral or mosque are readily admitted as "sacred" spaces for religious practitioners, of course, but a secular nation's capitol and its seat of democratic governance is also said to be "sacred." Human rights and freedoms are likewise "sacred" in this modern sense. But so, too, is a tribe's totem or initiation rite "sacred" to its members. For those sensitive to ecology and the dynamic systems that render all life possible, Nature and the biosphere are held as "sacred."

When threatened or attacked, these sites of the sacred will be adamantly defended by their acolytes like white blood cells warding off disease. That is because, without the sacred, the psycho-social superorganism would break down.[120] A tribe

depends on its totems, religious communities on their Gods, modern secular nation states on their democratic ideals and institutions, etc. They are what ground and orient such societies at their varying levels of size and complexity. They are what keep them from entropic dissolution as well as what propel them towards still greater integrity and scope.

As meaningful information, the sacred also complexifies through Symbolic Learning. In fact, its informational richness increases in tandem with the society's causal power, as it both permits and demands greater energetic input to sustain itself in the face of entropic dissolution. In this sense, the sacred itself is a kind of dissipative structure, a self-organizing informational configuration far from equilibrium in continual exchange with the environment. One might call this a *thermodynamic theory of the sacred.* Like all dissipative structures, the sacred complexifies by means of learning, and on the plane of Culture, this learning is accomplished by individual members of society.

As human minds complexify through learning, allowing more information to be related and coordinated, meaningful knowledge increases and the comprehensive scope and coherence of the sacred expands accordingly. The world model extends its horizons; the causal connections between phenomena grow more refined; the profundity of explanatory patterns deepens; the resolution of thought becomes sharper, more nuanced, more fine-grained. Entirely new superlatives of normativity come into view. What is best, highest, greatest, most important, most true, most real are articulated with more clarity and inclusivity. In the process, the prevailing sense of

the *ultimate* becomes successively contingent relative to something more ultimate still, and the sacred outgrows itself.

Given how important and, indeed, meaningful this process is, we may wish to appreciate *the learning process itself* as sacred—or (because it is what gives rise to sacred content) perhaps even as a sort of *meta-sacred* phenomenon. Learning is the means by which the sacred emerges and develops. After all, it is particularly in moments of breakthrough, of transcendence, of insight, that the sacred is experientially encountered. This is how new depths of meaning get disclosed. Such instances of adaptive learning mark advances into the unknown—a relatively chaotic realm of possibility whose increased entropy does not ultimately *disrupt* coherence but actually *enhances* it by the disclosure of still greater structure operating with greater degrees of freedom. But that, you will note, *is what complexification is.* We might suggest, then, that learning is the archetypal act of transcendence. It is what complexifies the mind, permitting an updating of the world model according to which cultural meaning is (re)assessed.

This meta-sacred act transcends any contextual sacred (e.g., a totem, a church, etc.) for it is the very pattern by which the sacred manifests and develops. Indeed, it is perhaps only because every instance of the sacred is itself a product of this general process that instances of the sacred can serve as a reliable canon by which to interpret novel information. As we shall see, aspects of this meta-sacred process wind up imprinted in specific sacred contents across the complexification process. Some things register as meaningful precisely because they gesture towards the meaning-developing process itself.[121]

We can thus appreciate that something deeper than the sacred underlies the development of the sacred—something more important than any local realization of importance. Predictive processing models of cognition may be indicative of this, as they suggest that it is *updating* our sense of meaning that is actually most meaningful to us. We operate off of prior information to make predictions, and if that information proves reliable, it is meaningful for us in the sense of maintaining our viability. But successful predictions are boring—by design. All is going very well viability-wise if we experience very little surprise. It means the model is holding. By contrast, it is when we are confronted by new information which may stand to profoundly upgrade the quality of our priors that we become *engrossed*, *transfixed*, *curious*, *amazed*. As we will see, awe is itself a product of the learning process, a state of wonder aroused by the attempt to integrate a surplus of potentially meaningful information.[122]

The experience of meaning as profound relevance, as salience, as significance—this arises not so much from any established sacred per se, but from the meta-sacred call to learn, to complexify, to transcend, and update one's conception of the sacred thereby. New forms of the sacred always lie on the other side of awe. Awe is the meta-sacred response to information we have not yet processed for its sacrality. It is the mind's reflex before the invitation to transcendence. It is the liminal state beyond one's old concept of the sacred and before a new one. It is the blissful open receptivity to an upgrade for one's meaning-making. It is the condition in which one meets such benevolent chaos as reorganizes you for the better—an incomprehensible fullness

appreciated by the mystics as the Mystery. All learning strives to render it known, and all learning ultimately falls short. What it produces is a series of "ultimates" progressively approximating more and more of that abyss of meaning.[123] These are the faces of God as they get redrawn through the advances in Symbolic Learning.

The evolution of meaning has thus been characterized by a succession of increasingly comprehensive and complex articulations of the sacred, to which human minds have transcended by reaching ever higher levels of informational integration and hierarchical coordination. Like other informational structures moving further from equilibrium, the configurational organization of the sacred has undergone a series of phase shifts across human cultural history, a punctuated equilibrium of different states reaching new degrees of complexity as more energy enters the system and organizes what it flows through. The substrate of this complex adaptive system is not some tangible solid or liquid, but the psycho-cultural system of information processing known as human society itself. The attractors to which it is drawn therefore correlate to the ways human minds learn to process symbolic information at higher levels of complexity. Meaning and the sacred emerge and evolve out of the learning mind's distinct Umwelts as it progressively renders more of the Mystery known.

In this way, knowledge and value extend in a graded continuum throughout the learning process of complexification. Feuerbach's insight that "each plant would declare its own flower the most beautiful" captures this transjective nature of relative good matching relative

understanding. But, being himself a theologian, Feuerbach took the idea even further, seeing that every entity must then have its own sort of guiding ideal, its own sacred, shaped by (what we can now call) the normative conditions of its entity–field relationship.

"If God were an object to the bird," he writes, "he would be a winged being: the bird knows nothing higher, nothing more blissful, than the winged condition."[124] There is, we might say, a God for every Umwelt, since meaning is adjudicated according to the information linking systems to their environment. This line of thought leads Feuerbach to posit that the God-concept itself is always but a given entity's intimation of ultimate knowledge:

> God is thy highest idea, the supreme effort of thy understanding, thy highest power of thought. …God is what the understanding thinks as the highest. But in what I perceive to be essential is revealed the nature of my understanding, is shown the power of my thinking faculty. …As thou thinkest God, such is thy thought;—the measure of thy God is the measure of thy understanding.[125]

As we shall see, though, a word like "God" is probably better reserved in reference to a particular cultural Umwelt's conception of the sacred. "God," at least as an anthropomorphic, supernatural being, is but one shape assumed by the sacred in cultural history, a moment in the complexification of meaning.[126] However, those who use the word "God" to refer to more abstract notions, such as Paul

Tillich's "ultimate concern," would be employing the term more as I am using "the sacred." I myself will sometimes speak of God in this way: not as any given moment in the evolution of the sacred, but as the enduring entity being evolved.

Indeed, over the following pages, I mean to show how God in this broader sense has evolved, complexifying through the learning process of cultural evolution. Once we are familiar with the distinct patterns of ontogenetic learning (which the next book will explore) and understand why and how they shape entire collective justification systems (discussed over the two subsequent books), we will be ready to track the evolution of meaning, value, and the sacred as they developed across human cultural history.

After 13.8 billion years of cosmic complexification, during which meaning itself was slowly growing more intricate and profound, the sacred at last emerged, full force, in human culture, where God awakened in the mind of mankind. From Structural to Genetic to Cognitive to Symbolic Learning, we come at last to the flower of evolution and consider the unfolding of sacred meaning in the human cultural domain. That will be the focus of our attention going forward.[127]

NOTES

---

[1] For a recent meta-analysis of the scientific literature on the link between meaning and well-being, see Katarzyna Czekierda et al., "Meaning in Life and Physical Health: Systematic Review and Meta-Analysis," *Health Psychology Review* 11, no. 4 (2017), https://doi.org/10.1080/17437199.2017.1327325.

[2] See, for instance, the article by behavioral scientist and meaning researcher Clay Routledge, "Suicides Have Increased: Is This an Existential Crisis?," *New York Times* 2018, https://www.nytimes.com/2018/06/23/opinion/sunday/suicide-rate-existential-crisis.html. Routledge notes that the 25% increase in suicides since 1999 demands explanation beyond a frayed mental healthcare system. He writes: "As a behavioral scientist who studies basic psychological needs, including the need for meaning, I am convinced that our nation's suicide crisis is in part a crisis of meaninglessness."

Since Routledge's piece in 2018, things have largely gotten worse for meaning and wellbeing in contemporary culture. Depression, anxiety, and self-harm are on the rise. For data and analysis, see Jonathan Haidt, *The Anxious Generation: How the Great Rewiring of Childhood Is Causing an Epidemic of Mental Illness* (New York: Penguin Press, 2024).

[3] Think, for example, of a waterwheel. Its power comes from a gradient. Water upstream naturally flows downstream, following the "path of least resistance" via energy minimization as it seeks "equilibrium" at sea level. Such an uneven gradient causes a movement whose energy can be harnessed as power to do work. Likewise, electrical power is generated through the gradients of positive vs. negative charge. One can do work using gradients in temperature, pressure, etc. In all cases, this is because gradients are inherently unstable in a cosmos fundamentally geared towards uniformity. The larger the gradient, the more work it can do when collapsed. A rubber band stretched 10 inches will spring back more forcefully than one stretched only 5. It is precisely in the universe's primal proclivity to "even things out," then, that energy does work.

[4] **Source:** https://commons.wikimedia.org/wiki/File:Diffusion.svg.

[5] The seminal paper is Claude Shannon, "A Mathematical Theory of Communication," *Bell System Technical Journal* 27, no. 3 (1948).

[6] See David Krakauer et al., "The Information Theory of Individuality," *Theory in Biosciences* 139, no. 2 (2020), https://doi.org/10.1007/s12064-020-00313-7. The authors note that

> without a rigorous definition of both the environment and the agent it is difficult to speak consistently of individuals. This is analogous to figure-ground separation in gestalt psychology or computer vision. The background of an image carries as much if not more information than the object, and the challenge is to separate the two rather than assume that they are already distinct and independent. (211)

We can thus understand entities in contexts (i.e., systems in environments) as being mutually defined in terms of the other.

Krakauer and his colleagues offer a mathematical formalization of this relationship for adaptive entities in terms of information theory. What is fundamental for individuality, they argue,

> is the idea that information can be propagated forward through time, meaning that uncertainty is reduced over time. In this way…we suggest individuality is a natural extension of the ideas of Boltzmann and Von Neumann, and as such has foundations in statistical mechanics and thermodynamics, which consider the conditions required for a persistently ordered states." (214)

The figure-ground distinction is a helpful way of appreciating the system–environment relationship through the lens of information and entropy. Here, equilibrium can be compared to homogenous uniformity with no contrast between figure or ground:

The existence of an entity, then, is predicated on drawing such a distinction, cleaving uniformity into some duality of figure and ground, thing and context:

Such differentiation creates system and environment simultaneously. To be a thing, then, is always to be a thing in context, since identity is defined relative to the "negative space" that distinguishes it.

Moreover, such separation is what "informs" the entity, as the maximal entropy of mixed, homogenous equilibrium (statistically the most likely configuration) becomes reduced through the segregation into thing and not-thing (a less likely configuration), yielding an increase in information.

Appreciating this distinction as a gradient, we can see that its maintenance requires energy, since the Second Law entails the dissolution of gradients toward equilibrium absent a source of energy that can do the work of keeping them apart. The more informationally rich the entity, the more energy will be required to keep it far from equilibrium. Entities, then, can be understood as information propagated through time, employing the free energy dynamics of their contexts to maintain themselves in the face of entropy.

[7] **Source:** Ibid., Figure 1.

[8] We are here dealing with some fundamental philosophical categories whose relationships to physical reality are still being clarified in the philosophy of science and the development of new scientific paradigms. Specifically, I am referring to the issue of identity through definition, and thus distinction, differentiation. For a thing to be it must *not* be something else. The figure-ground analogy creates a defining division, a difference that *matters* (in every sense of the term), an in-forming difference. As Gregory Bateson famously put it, information is "a difference that makes a difference." Without difference, there is only the homogeneity of equilibrium.

What generates distinction are *constraints*: limitations of the possibility space of equilibrium that allow for difference. As Alicia Juarrero writes in *Context Changes Everything: How Constraints Create Coherence* (Cambridge: The MIT Press, 2023):

> At thermal equilibrium, nothing is more probable than anything else; since there is no organization, neither is anything qualitatively different from anything else. Distinctions and thermal nonequilibrium—that is, inhomogeneities in energy flow—are prerequisites for coherence-making, information, communication, and work. *Context-independent constraints* take conditions away from equilibrium. They render conditions, events, and processes that were equally likely no longer *equiprobable*... They establish boundaries of uneven possibility landscapes (like fields) within which energy can flow and other constraints can emerge. (50)

Information, then, is differentiation generated by limiting the equiprobable outcomes of equilibrium; that is the difference that makes a difference. Constraints differentiate phenomena, and thus establish informational entities relative to more entropic environments. By contrast, equilibrium is literally "no-thing." Juarrero writes:

> Once reached, thermal equilibrium lasts, but there is *no-thing* that persists at thermal equilibrium… White noise is no-thing; equiprobability is no-macrostate… [A]t thermal equilibrium there are no constraints and consequently no stored information that is qualitatively other than the aggregate or sum of the equally likely *microstates*. (131)

Such a framing necessarily reorients how we think about, well, every-*thing*, suggesting a metaphysical reconceptualization of reality itself. In *Essential Difference: Toward a Metaphysics of Emergence* (Albany: State University of New York Press, 2012), James Blachowicz attempts such a philosophical reorientation, writing:

> It is the essential material differences among physical entities – and not determinations that establish their commonality – that constitute them. In terms of constraint interpretation, these differences are coincidental with constitutive constraint itself. (224)

The mistake of reductionist accounts is that they lose these "essential differences" precisely in the attempt to completely causally explain them relative to something more common and basic (and thus with fewer differences/constraints). In reality, biology cannot be reduced to physics, however, because biological phenomena exhibit constraints over and above physical ones. That is what gives them their unique properties and makes possible their unique forms of behavior.

> Qualitative differences between physical entities thus emerge as quantitative differences between levels of individuation and differentiation of physical entities. This is the basis for an ontological hierarchy of physical entities.
>
> Physical entities are structures of structures. A biological organism is ultimately a complex higher-level constraint on the degrees of freedom of its component molecules; each molecule in turn is a constraint on the degrees of freedom of its components. Higher levels are more inclusive than lower levels, involving a

> greater number of constitutive constraints; in this respect, higher-levels provide a *quantitative* augmentation of what is found in lower levels. Yet we must also consider that such higher levels (understood in this nested way) represent *qualitative* enhancements in the nature of individuation and differentiation as well – demonstrating what primitive subatomic matter is ultimately capable of. At each level, these constraints constitute new matter, and are not simply non-constitutive properties of bottom-level physical "simples." (258)

(For a deeper philosophical presentation of these ideas, see John Vervaeke's keynote address to the 2023 Consilience Conference: https://www.youtube.com/watch?v=vazO36OnGKI.)

Because constraints are the basis of information, we can say that biological organisms are more informationally rich than physical objects; that is, they have more limitations on their possibility space than physical objects do—psychological beings, even more limitations; and so on. The more complex the entity, the more constraints are at work distinguishing it from equiprobable equilibrium, meaning more information defines it.

Despite the connotations of the term, though, constraints are not purely restrictive but *enable* new possibilities precisely by narrowing the field of possibilities. For instance, by constraining the path of a river through a dam, power can be harnessed to do all sorts of creative work. By constraining the aperture of my garden hose as it flows, I can get a more forceful jet that can more effectively clean things and push matter around. The complex network of neurons in my brain allows me to be conscious and aware—an incredibly improbable configuration demanding many constraints. By this logic, then, more constraints actually entail more powerful and autonomous entities, as specific limitations create unique affordances. In cultural and biological terms, for instance, Juarrero observes:

> New rights and opportunities appear concomitantly with new responsibilities and constraints. Without the emergent affordances offered by legal and judicial platforms through their enabling and governing constraints, individuals cannot be spouses, mayors, or conscientious objectors, for example. …As Locke, Hobbes, and Rousseau well understood, such roles (and authorizing powers) are unavailable to a motley group of people prior to coalescing as a social organization of a particular sort (organization being defined in terms of those interlocking constraints that establish and stabilize coherence).

> Similarly, genetic material captured in an amino acid, amino acids captured in a protein, and mitochondria captured in cytoplasm have fewer degrees of freedom than as isolated biochemicals. Amino acids are constrained components of the coordination dynamic that constitutes a protein. The wider set of interdependencies (which we call protein), however, acquires properties and functionalities the amino acids individually lack. (86)

In short, through increases in differentiation, and the integration of that differentiation into higher-order entities, things *complexify*, gaining emergent properties in the process. With each higher-order integration, then, there will be more information defining that entity—the essential difference that makes a difference.

[9] Artemy Kolchinsky and David H. Wolpert, "Semantic Information, Autonomous Agency and Non-Equilibrium Statistical Physics," *Interface Focus* 8, no. 6 (2018), https://doi.org/10.1098/rsfs.2018.0041.

[10] Ibid., 1.

[11] Ibid., 2. To assess and measure the causal power of such information, Kolchinsky and Wolpert draw on the causal calculus of Judea Pearl. For a non-technical introduction to this framework, see Judea Pearl and Dana Mackenzie, *The Book of Why: The New Science of Cause and Effect* (New York: Basic Books, 2018).

[12] Kolchinsky and Wolpert, "Semantic Information," 1.

[13] Ibid., 9.

[14] Ibid., 1. Recalling the system–environment schematic of Figure 1.2, we could say that, for the information constituting a system (S_n) to be accurately transmitted to the system in a future state (S_{n+1}), it must relate to information from the environment (E) in a manner that ensures its resistance to entropy.

[15] Carlo Rovelli, "Meaning and Intentionality = Information + Evolution," in *Wandering Towards a Goal: How Can Mindless Mathematical Laws Give Rise to Aims and Intention?*, eds. Anthony Aguirre, Brendan Foster, and Zeeya Merali (Cham: Springer International Publishing, 2018), 17–28.

[16] Bobby Azarian, *The Romance of Reality: How the Universe Organizes Itself to Create Life, Consciousness, and Cosmic Complexity* (Dallas, TX: BenBella Books, Inc., 2022), 127; italics original.

[17] Ibid., 139. Or, as he puts it elsewhere: "All evolutionary processes are learning processes, and all forms of learning are evolutionary processes" (94).

[18] In essence, I believe the concept of "meaningful information" grounded in the energy-information dynamics of the physical universe allow us to frame meaning and value in terms of first principles. By definition, first principles are irreducible to more fundamental causes. The Second Law is the classic example from physics. It cannot be further explained in terms of anything else. We know that the universe naturally homogenizes gradients, but we cannot say *why* this is so in a manner that explains this phenomenon on the basis of anything more elemental. Hence its status as a Law of nature.

Explanations grounded in first principles have (provisionally) complete causal closure. Unlike most phenomena, whose causes are themselves contingent on still other, prior causes, first principles offer, if not a foundation, then a horizon or limit for justified explanation. When we ask what and why something is, we are naturally led through a regressive succession of contingent causes. It is a reality common to both inquisitive toddlers as well as scientists and philosophers that the answer to one Why is just the basis for another question. (Indeed, the importance of justification dynamics to the development of human knowledge will itself be an important topic later in our discussion.)

Faced with the issue of infinite regression, Aristotle famously offered the ultimate backstop of a *causa sui*, a "self-caused cause," an unmoved mover. By equating this ultimate cause with a thinking mind (per Greek philosophy) and personal God (per Judeo-Christian religion), Western thought before modernity tended to assert meaning and purpose as cosmic primitives. Not just the causal chain of events but also the origin and basis of value was laid at God's door as the metaphysically necessary guarantor of significance. This had the effect of framing value as something given, inherent in Being and beings. With a personal God as the eternal source of goodness, value was considered absolute, intrinsic to nature.

This framing profoundly shifted with the rise and spread of modern thought. Through incisive scrutiny of material and efficient causes alone, science occasioned a widening distinction between objective and subjective phenomena. Cartesian rationality helped move the locus of meaning and value from things to minds, which were still held to be divine. Eventually, though, divinity would evaporate altogether from rational mechanistic accounts, leaving meaning and value as mere subjective *experiences* playing out in an otherwise objective world which itself lacked any inherent meaning.

Since then, modern and postmodern thinkers have grappled with the implications of such a subjectivist account of meaning (e.g., existentialism,

radical social constructivism, etc.)—to unsatisfactory conclusions. The notion that meaning and value are simply "made up" and arbitrary proves neither rationally nor psychologically satisfactory. But neither can we justify the glaringly biocentric if not anthropocentric notion that a specific (and quite human-friendly) sense of purpose ("final cause") is somehow fundamental, context-independent, and absolute.

The present argument offers an intervention into this philosophical impasse between absolutistic and relativistic conceptions of meaning and value. The nature of meaning can indeed be grounded—not in the inherent nature or essence of things but in the intrinsic conditions that obtain (and *must* obtain) for a thing to exist in context. All things are things-in-context. As soon as a thing exists, its meaning is defined by its environmental conditions of existence (cf. the figure-ground distinction in note 6). To continue to be is a necessary goal and precondition for being. Being, then, begets meaning. Meaning *is* the existential imperative.

Value emerges with entities, then, as value is always value *for*, and if there is nothing *for* then there is no value. Value is not intrinsic *to* existence, then, but *for* existence. For something to be and propagate its information through time creates the intrinsic requirements of continuing to be, whose conditions for meeting define the value-sphere of the entity in question. If they are not met, the entity no longer exists. So, you might say, nature selects for meaning. By virtue of being, there is mutual information linking system and environment to further being.

This is where the famous "is/ought distinction" ultimately bottoms out, beyond which the distinction ceases to make sense, since here being and normativity co-form one another. Entities which do not capitalize on meaningful information cease to exist. Those that do continue within a field of meaning.

As we shall see, these informational relationships naturally complexify over time, creating more informationally rich entities requiring more energy to sustain themselves. If a complex being feels no *ought* for its own *is*, then it will cease to be. If it does, then it persists, and so meaning persists—and keeps evolving. This basic insight is captured by biologist Andreas Weber's 'First Law of Desire': "Everything that lives wants more of life. Organisms are beings whose own existence means something to them" (see Andreas Weber, *The Biology of Wonder: Aliveness, Feeling, and the Metamorphosis of Science* (Gabriola, BC: New Society Publishers, 2016), 3–4, 28–29). As we shall see, though, this normativity of being extends even to the pre-biological level of existence as well. "Desire" itself is a more complex instantiation of a more fundamental reality.

Such a framing provides a first principles basis for meaning in a manner that avoids the pitfalls of premodern absolutist notions and (post)modern relativist conceptions (whose conceptualization itself represents an evolution of meaning, as we shall see). Being is inherently relative through the co-defining relationship of entity and environment; but it is precisely this *relationality* that offers the *absolute* basis for meaning. Meaning is an absolute relationality, a universal contextuality, a fundamental contingency. Meaning is *causa sui.*

[19] When it comes to informational complexity, consider two entities in an environment: a rock and a cell. Both must faithfully transmit their information to a future state if they are to persist. But the further-from-equilibrium cell has considerably more information to transmit than does the rock.

One might use *Kolmogorov complexity* to represent this reality, which is a generalized version of Shannon information theory, also called "algorithmic complexity" because it refers to the length of the shortest "algorithm" or "code" needed to communicate a string of information. For instance, for two messages of 18 lowercase letters, *ababababababababab* and *sjmgxwoslgjauyqhrjxmbn*, one can be expressed as "write *ab* 9 times" (which consists of only 16 characters) while the other has no obvious simple description besides just writing the string itself. The second message is thus more "irreducibly complex" while the first is more predictable ("compressible") and thus simpler. In thermodynamic terms, the second code will require more computing power to run as it is more informationally rich (which is precisely why "compressing" files to a .zip file is a thermodynamically economic application of algorithmic complexity).

If we liken the simple information of a rock to being more like message 1 and the more complex information of a cell to being more like message 2, we get a sense of the complexity spectrum of different entities in informational terms. (Recalling the framing of *essential differences* and *constraint* explored in endnote 8, we might note that message 1 has fewer differences than message 2.) Complexity emerges in the dynamic space between ordered uniformity and chaotic randomness.

[20] Eric Chaisson, *Cosmic Evolution: The Rise of Complexity in Nature* (Cambridge: Harvard University Press, 2001), 131.

[21] **Source:** Ibid., Figures 28 and 27b.

[22] Free energy rate density measures the amount of energy that flows through a system within a given amount of time and mass—so, for instance, how many ergs or joules of energy are flowing per second per gram of a given system. Chaisson notes that

> open-system, non-equilibrium thermodynamics is not concerned with the absolute value of a structure's total free energy as much as with its free energy density. After all, a galaxy clearly has more energy than a cell, but of course galaxies also have greater sizes and masses; it is the organized energy *density* that best characterizes the degree of order or complexity in any system… In fact, what is more important is the *rate* at which free energy transits a complex system of given mass; accordingly, all systems can be compared on a fair and level spectrum. In other words, a most useful quantity used to specify operationally the order and organization of any system is the free energy rate density…expressed in units of energy per time per mass and denoted by the symbol Φ_m. (Ibid., 133-134.)

Chaisson has made the case for seeing this metric as the best universal measure of complexity in the cosmos. "Better metrics than energy rate density," he writes, "may well describe each of the individual systems within the realms of physical, biological, and cultural evolution, but no other single metric seems capable of uniformly describing them all" ("Using Complexity Science to Search for Unity in the Natural Sciences," in *Complexity and the Arrow of Time*, ed. Charles H. Lineweaver, Paul Davies, and Michael Ruse (Cambridge: Cambridge University Press, 2013), 73). See also "Energy Rate Density as a Complexity Metric and Evolutionary Driver," *Complexity* 16 (2011) and "Energy Rate Density II: Probing Further a New Complexity Metric," *Complexity* 17 (2011).

Chaisson is right that other complexity metrics become particularly valuable for the specific planes of complexity beyond physics. Genomic complexity (Life) might be amenable to algorithmic complexity metrics, for instance. And, as we shall see below, metrics for integrated information provide important insights about neuronal complexity (Mind). Most important for our purposes, however, will be the metric for symbolic semantic complexity (Culture) called "hierarchical complexity." Still, Chaisson is right that free energy rate density offers the best base metric, since all of these metrics for higher planes of complexity are predicated on increasing energy flows generating more complex structures capable of doing novel kinds of information processing—as Chaisson's calculations show.

[23] Eric Chaisson, "The Natural Science Underlying Big History," *Scientific World Journal* 2014 (2014): 5, https://doi.org/10.1155/2014/384912.

[24] Chaisson, *Cosmic Evolution*, 133.

---

[25] Azarian, *The Romance of Reality*, 99.

[26] **Source:** Ibid., Figure 2.

[27] This directionality of cosmic evolution towards more energy coursing more densely through more complex structures is nicely complemented by frameworks like Adrian Bejan's "constructal law," which states: "For a finite-size flow system to persist in time (to live), its configuration must evolve in such a way that provides easier access to the currents that flow through it" (Adrian Bejan and J. Peder Zane, *Design in Nature: How the Constructal Law Governs Evolution in Biology, Physics, Technology, and Social Organization* (New York: Doubleday, 2012), 3). If equilibrium is a static homogeneity due to there being no gradients, and thus no movement, then the emergence of design in nature owes precisely to the success of dynamic *flow* channels. "In all instances," writes Bejan, "design emerges because things flow better with configuration" (13).

Indeed, this is the same logic behind "dissipative structures," discussed below. Stated briefly, gradients are collapsed more efficiently through orderly structures. In a universe governed by the Second Law, and thus necessarily driven to equilibrate gradients and produce entropy, *order emerges spontaneously to facilitate the drive towards disorder*. Ironically, then, design in nature is the natural byproduct of a universe ostensibly driven to erode design. To better facilitate flow, and thus the generation of entropy, more and more complex flow configurations will emerge to dissipate free energy faster.

Such dynamics, says Bejan, mean there is in fact a direction to evolution:

> The constructal law...challenges another idea that has become dogma since Darwin—that there is no overarching direction to evolution. Proponents of that view claim that adaptations make species better able to survive, but they never explain why these changes should occur and what they mean by "better." The closest they come is through a piece of circular logic that says: A change is better if it aids survival; any change that aids survival is better. The constructal law, by contrast, predicts that evolution should occur because of the tendency of all flow systems to generate better and better designs for the currents that flow through them. It expresses the meaning of "better" in unambiguous physics terms—change that facilitates faster, easier movement. ...[N]ot only do river basis and forests improve in time but so do biological creatures—the rise of species from single-cell organisms to fish, birds, and humans is the story of better, more efficient flow of animal mass on the

> landscape. In big history, all these designs have emerged because they enhance the movement, mixing, and churning of energy and mass on the planet. (24)

Indeed, Bejan even goes to far as to describe this as the *meaning* of evolution itself, saying: "In the big picture—all the rivers, all the trees, all the animals on Earth—the driving force behind the evolution of everything that flows is the generation of shape and structure to move more easily. This is why we can say that this unintentional tendency has a *purpose*" (35; italics original).

While it's certainly true that the direction of cosmic evolution is towards greater flows of energy within more complex structures, as both Chaisson and Bejan rightly note, is creating entropy actually evolution's "*purpose*"? This idea has had its proponents, from Dorion Sagan and Jessica Hope Whiteside's "Gradient Reduction Theory" to, more recently, transhumanist sympathizers of "Effective Accelerationism." Unfortunately, such a perspective puts the proverbial cart before the horse in a profound way. To assert that entropy is our purpose is like saying the purpose of a car is to burn fuel or the purpose of a laptop is to generate heat. In fact, energy consumption and structural complexity are but byproducts of that *other* thing that's exponentially increasing: *information.*

While complexification may start out at a mechanistic level best explained by entropy production in accordance with the Second Law, the process itself undergoes multiple phase transitions as complexification unfolds, whose emergent properties successively reveal a different significance to such dynamics. Cosmic evolution uses energy to fuel an ongoing *learning process.* The result is net entropic increase in accordance with the Second Law, but the direction is towards more information, more knowledge, more consciousness, *more meaning.*

Azarian understands this. He writes:

> Your literal purpose in life is not to create as much disorder (entropy) as possible—as some have interpreted the thermodynamic narrative—but to acquire knowledge that preserves and expands the order constructed by life. However, the collective result of agents who behave in this way is a biosphere that grows increasingly diverse, complex, and computationally powerful over time, such that the rate of total entropy production is maximized over a time window that extends into the future. (*The Romance of Reality*, 140)

Getting this point right is essential, I believe, if we are to fully appreciate the implications of complexification. As we shall see, the story of emergent complexity and the universal learning process itself represents an evolution of meaning at the level of worldview. Increasingly, people are seeing the thermodynamic writing on the wall, but the meaning they draw from it has differed. The argument of this book series is that this process has a profoundly meaningful and spiritually fulfilling interpretation—one entirely missed if we mistake means for ends. Yes, the cosmos has a direction towards increasing entropy production through complexification—but seeing this as the purpose of evolution is literally to misinterpret our goal as a waste product. The reality is much more interesting, compelling, and sublime.

[28] Seth Lloyd, "The Computational Universe," in *Information and the Nature of Reality*, ed. Paul Davies (Cambridge: Cambridge University Press, 2014), 125.

[29] Gregg Henriques, *A New Synthesis for Solving the Problem of Psychology: Addressing the Enlightenment Gap*, Palgrave Studies in the Theory and History of Psychology (Cham: Palgrave Macmillan, 2023), 282.

[30] The idea of emergence has long been central to complexity science. Though, like the term "complexity" itself, there are ongoing debates about its nature. Essentially, emergence refers to those behaviors of complex systems that are not evidenced at the level of the component parts but only at the level of their composite whole. For instance, the flight behavior of a single starling looks very different from the flight behavior of a whole murmuration of them together. The murmuration is a network of starlings which exhibits emergent properties that are not present in the individual parts but only emerge when many starlings operate together. Another common example is the behavior of an ant colony, which is truly functional only at the level of a superorganism. In isolation, individual ants are not independently viable. In the context of the colony, however, the collective behavior of the multitude creates an emergent whole that *is* viable, and includes complex subsystems like food production, defense, etc. The ant colony thus possesses emergent features that its constituent parts do not.

Understanding how emergence works has been an ongoing discussion in the field of complexity. In *Complexity: A Guided Tour* (Oxford: Oxford University Press, 2009), Melanie Mitchell defines a complex system as

> a system in which large networks of components with no central control and simple rules of operation give rise to complex collective

> behavior, sophisticated information processing, and adaptation via learning or evolution. …Systems in which organized behavior arises without an internal or external controller or leader are sometimes called *self-organizing*. Since simple rules produce complex behavior in hard-to-predict ways, the macroscopic behavior of such systems is sometimes called *emergent*. Here is an alternative definition of a complex system: *a system that exhibits nontrivial emergent and self-organizing behaviors*. The central question of the sciences of complexity is how this emergent self-organized behavior comes about. (13)

Differing accounts of emergence have hinged on whether it is ontological or merely epistemological, "strong" or "weak." If "more is different," as the saying goes, how and why is this the case? Where do the new governing principles come from if not "bottom up" mechanisms that still allow for fully reductionistic explanation?

Emergent accounts are essentially the opposite of reductionistic ones. While reductionism analytically seeks causal mechanisms at the lowest level of constituent parts (i.e., psychology is causally reducible to biology, which is reducible to chemistry, which is reducible to physics), emergence posits that causal power can pertain at the level of composite wholes (cf. endnote 8). Complex systems, it is argued, can causally influence their constituent parts—actually controlling them "top down" and not simply being controlled *by* them "bottom up." In such a view, psychology would not be causally reducible to physics; rather, minds actually agentially influence their material substrate. Minds can control and direct their bodies and not simply serve as passive mechanisms of reductionistically deterministic atoms (as so-called "epiphenomenal" accounts of consciousness posit).

Challenges to such conceptions of "top down" causation (cf. Jaegwon Kim, *Mind in a Physical World: An Essay on the Mind-Body Problem and Mental Causation* (Cambridge.: MIT Press, 1998)) have been roundly countered by multiple lines of confirming argumentation from systems thinkers. Their formal arguments employ information theory and far-from-equilibrium thermodynamics to make the case that emergent properties do indeed have causal power such that macroscale phenomena cannot simply be reduced to their constituent parts. Applying Judea Pearl's causal calculus (cf. endnote 11) to assess the causal power of micro- vs. macro-scale information, Erik Hoel, for instance, has demonstrated that macro-scale entities do indeed exhibit causal power over and above the micro-scale constituent parts. He defines such "causal

emergence" as occurring "when a macroscale contains more information and does more causal work than its underlying microscale" ("Agent Above, Atom Below: How Agents Causally Emerge from Their Underlying Microphysics," in *Wandering Towards a Goal: How Can Mindless Mathematical Laws Give Rise to Aims and Intention?*, eds. Anthony Aguirre, Brendan Foster, and Zeeya Merali (Cham: Springer International Publishing, 2018), 66). Because wholes are systems of many parts, parts can be continuously renewed and replaced even as the whole itself persists. Wholes exhibit "multiple-realizability," meaning many different micro-scale configurations can all produce the same macro-scale phenomenon. Multiple-realizability gives macro-scale wholes a level of redundancy that makes them more causally potent. Hoel writes:

> Quantifying this micro-to-macro change can be done using information theory. This is possible because causal structure is a matrix that transforms past states into future states; it thus can be conceived of as an information channel. In this view, individual states are like messages and interventions are like the act of sending them over the channel. We can then ask how much each state (or intervention) reduces the uncertainty about the future of the system by measuring the mutual information between some intervention distribution and its resultant effect distribution: $I(ID;ED)$. It's an equation that quantifies Gregory Bateson's influential definition of information as "a difference that makes a difference." (Ibid., 67-68)

That is, macro-scale entities can possess greater causal power than micro-scale entities because their redundancy makes them less susceptible to noise in the same way that "error correction" works in information theory. Like the Ship of Theseus, the whole robustly persists even as its parts undergo continual entropic breakdown and repair. More is indeed different, allowing for the emergence of causal powers not reducible to an entity's constituent parts.

Hoel posits that the scales of study in science (e.g., chemistry, biology, psychology, etc.), far from all being reducible to elemental physics, actually represent emergent zones of unique causal significance. As he puts it,

> changes in scale are a form of channel coding, and by encoding up at macroscales causal relationships can error-correct, do more work, generate more information, and offer a better model choice for experimenters. It may be that scientists *en masse* implicitly choose

> certain scales for this reason, and the creation of a hierarchy of science is truly following Plato's suggestion to "carve nature at its joints." (Ibid., 72)

See his chapter "How Science Got Its Scale" in *The World Behind the World: Consciousness, Free Will, and the Limits of Science* (New York: Avid Reader Press, 2023), 161–194.

Such an idea fits very well with the Tree of Knowledge's presentation of the different emergent levels of reality. Physics is the scientific field for assessing the causal behavior of Matter, biology for Life, psychology for Mind, and the social sciences (e.g., sociology, economics, etc.) for Culture. These realms are irreducibly complex, exhibiting emergent causal behavior over and above the realm upon which they supervene.

In his book *The Emergence of Everything: How the World Became Complex* (New York: Oxford University Press, 2002), biologist and complexity theorist Harold Morowitz lists 28 levels of emergent complexity that developed in the universe since the Big Bang, which can be nested between the ToK's four joint points for a more granular appreciation of cosmic evolution across scales:

[Matter]

1. The Primordium
2. Large-Scale Structure
3. Stars
4. The Elements
5. Solar Systems
6. Planets
7. The Geospheres

[Life]

8. The Biosphere
9. Prokaryotes
10. Eukaryotes
11. Multicellularity

[Mind]

12. Neurons
13. Two Animal Subkingdoms
14. Vertebrates
15. Fish
16. Amphibians

17. Reptiles
18. Mammals
19. Arboreal Mammals
20. Primates
21. Great Apes
22. Hominids
23. Toolmakers

[Culture]

24. Language
25. Agriculture
26. Technology and Urbanization
27. Philosophy
28. The Spiritual

For a more recent framework in this vein, see Tyler Volk, *Quarks to Culture: How We Came to Be* (New York: Columbia University Press, 2017); for Volk and Henriques's joint paper calling for a "Big History 2.0," see Gregg Henriques and Tyler Volk, "Toward a Big History 2.0: A Brief Position Paper," *Journal of Big History* VI, no. 3 (2023), https://doi.org/10.22339/jbh.v6i3.6301.

When we compare such emergent levels to Chaisson's calculations of free energy rate density, we see a direct overlap. With more information and causal power comes a greater need for energy to structure what it flows through. Indeed, it is through flows of energy that new structure emerges to better dissipate the energy. Azarian writes:

> Each emergence in the cosmic self-organization process is brought about by a phase transition that moves life ever further away from a state of thermodynamic equilibrium and total disorder. Through a nested series of such phase transitions—where functional things come together to make larger functional things, which come together to make even larger ones, and so on—adaptive complexity becomes better equipped to dominate the cosmos. That is, life acquires an organization that is increasingly hierarchical and integrated, and therefore more resilient and computationally powerful. (Azarian, *The Romance of Reality*, 164)

Such is the direction of evolution, as Bejan suggests: towards more complex designs that better facilitate flows of energy, and thus new

emergences. Fascinatingly, such a thermodynamic imperative is what drives a cosmic learning process unfolding across the growing stack of emergent levels, whereby more information gets processed by novel kinds of emergent agents, which acquire greater causal efficacy from Life to Mind to Culture.

[31] Gregg Henriques, *A New Unified Theory of Psychology* (New York: Springer, 2011), 14; emphasis added.

[32] **Source:** Henriques, *A New Synthesis*, Figure 6.3.

[33] **Source:** Adapted from ibid., Figure 10.11. The concepts of Mind^1 and Mind^3 will be dealt with below (see endnote 78).

[34] Ibid., 278.

[35] Azarian writes:

> Naturally, the complexity of the organism will evolve to match the complexity of the niche. More specifically, the complexity of the organism will correspond to the complexity of the environmental challenge the organism is confronted with, which starts off as a purely thermodynamic challenge. Some niches present a changing variety of challenges that must be adapted to while others present hardly any. As a result, some species don't need to evolve. Others, however, have niches that become increasingly complex over time. (*The Romance of Reality*, 134)

Likewise, Paul Davies and Michael Rule note in "What is Complexity? Is it Increasing?", their introduction to *Complexity and the Arrow of Time*: "When environments get more varied and complex, so too do the inhabitants. If the environment gets simpler (closer to equilibrium with fewer gradients) then the inhabitants get simpler as selection pressure is removed and the information in their genomes drifts away" (8). Indeed, as we shall see, it is the rising complexity of the environment that provides the source for the informational complexity of the genome through Genetic Learning. If there is less information in the environment, there is less information for the genome to learn through selective pruning.

[36] Wojciech Zurek, "Quantum Darwinism," *Nature Physics* 5 (2009).

[37] John Campbell, *Universal Darwinism: The Path of Knowledge* (Scotts Valley: Independently published, 2011), 33.

[38] Campbell frames this as an inferential learning process using the same Bayesian predictive processing framework as Karl Friston's free energy principle (discussed below). From the vast statistical possibility space of quantum states, most cancel out, and the ones that get "selected" are only always classical ones—the "only 'fit' quantum information [that]

can survive in its environment" (John Campbell, *The Knowing Universe* (Las Vegas: Independently published, 2021), 246). With this, the quantum information, now as classical reality, "updates" the quantum wave function to predict the outcome with probability 1. "This update," writes Campbell, "causes the wave function to evolve and better predict its environment" (*The Knowing Universe*, 266). In this way, quantum decoherence represents simply another instance of inferential processing—a *learning* process based on the propagation of information through an entropic environment. If this is true, everything in the known universe has meaning, technically speaking, at least at the level of the entity–field relationship of fundamental particles.

[39] Perhaps this should be framed as a form of "learning" *before* the Structural Learning of the Matter level (Quantum Learning?), since it is what leads to the emergence of material itself out of the Energy-Information implicate order. As Henriques writes: "In the parlance of modern physics, Matter can be thought of as the domain where Energy-Information quantum field fluctuations 'decohere' into measurable or observable entities that behave in accordance with classical mechanics on three dimensions of space and one dimension of time" (*A New Synthesis*, 122). If true, perhaps instead of "Quantum Gravity" as the theoretical basis of the first Joint Point in the Tree of Knowledge (currently just a "placeholder," ibid., 126), we should place "Quantum Darwinism."

[40] The idea of pre-biotic selection pressures leading to complexification of material structures is key to Assembly Theory (AT), presented in A. Sharma et al., "Assembly Theory Explains and Quantifies Selection and Evolution," *Nature* 622, no. 7982 (2023), https://doi.org/10.1038/s41586-023-06600-9. The authors state:

> AT provides a framework to unify descriptions of selection across physics and biology, with the potential to build a new physics that emerges in chemistry in which history and causal contingency through selection must start to play a prominent role in our descriptions of matter. (327)

Indeed, increasing complexity itself suggests increasing layers of selective pressure, they argue. "We conjecture that, the 'more assembled' an ensemble of objects, the more selection is required for it to come into existence" (326).

Evolution is based on variation, inheritance, and reproduction: Lewontin's "units of selection" (R. C. Lewontin, "The Units of Selection," *Annual Review of Ecology and Systematics* 1 (1970)). In biological

evolution, DNA provides the mechanism for inheritance, encoding information in the genome. But such replicable "memory" can be much simpler, allowing selection to occur even at the pre-biotic level. Sharma et al. write:

> By aiming to detect 'selection', we mean a process similar to selection in Darwinian evolution. We do not, however, model functional differences that selection might act on. Instead, we account only for the specificity of selection—that some objects are more likely to be used to make new things and some are less likely. The only functionality we want to detect or describe is in the memory of the process to generate the object, with examples including a metabolic reaction network or a genome. This allows the three Lewontin conditions for evolution to hold. (324)

Through this filtration process, nature "discovers" new assemblages that can stay assembled in their environments, which in turn get used as the basis for more complex assemblages. If evolution is an inferential process honing meaningful information to aid viability, then persistent assemblages represent successful learning at the purely material level.

[41] See Ilya Prigogine and Isabelle Stengers, *Order out of Chaos: Man's New Dialogue with Nature* (Boulder: Random House, 1984). The easiest dissipative structure to appreciate is probably the whirlpool gyre that forms in your sink or bathtub when you pull the stopper. When you do, a gradient is formed (i.e., tub full of water vs. drain empty of water). To dissipate this gradient most efficiently, a highly ordered spiral structure emerges: the whirlpool. In this way, at the simple level of Matter, order emerges spontaneously out of the natural tendency to homogenize differences.

Hurricanes are a much larger version of the same phenomenon, but there are many other examples of such dissipative structures ranging from convection currents to turbulence flows as well as various chemical reactions. Dissipative structures can be seen as a more complex product of Bejan's constructal law, which posits that flow channels will naturally evolve towards greater efficiency.

[42] Azarian, ever succinct, offers a wonderful summary of dissipative structures this way:

> Since nature loathes gradients, the effort put forth into reducing them is proportional to the magnitude of the gradient. When gradients become large enough and sufficient free energy becomes

> available, a cycling structure will *spontaneously emerge* to enable the gradient to dissipate more efficiently. Because the universe "wants" to produce entropy at the fastest possible rate, it will actually create organized systems that help it achieve its thermodynamic goal more efficiently. …Dissipative structures reveal that entropy production can be as much of a motor for order and complexity as it is for disorder. This discovery fundamentally changes the way we think about the second law: it is both the destroyer and creator of organized structure in nature. (*The Romance of Reality*, 36–37)

[43] With dissipative adaptation, entities actually assume specific configurations in response to the specific sorts of energy conditions of their environment, adapting to the environment in such a way as to most efficiently dissipate the free energy flowing into the system. For instance, it has been shown that shining a light on silver nanoparticles causes them to self-organize into more improbably orderly configurations that could capture and dissipate the specific frequency of the light most efficiently (S. Ito et al., "Selective Optical Assembly of Highly Uniform Nanoparticles by Doughnut-Shaped Beams," *Sci Rep* 3 (Oct 25 2013), https://doi.org/ 10.1038/srep03047). Similarly, it has been shown that introducing an electric current to conduction beads in oil causes them to self-organize into a dissipative structure with "wormlike motion," which continues as long as the energy flow persists (A. Belkin, A. Hubler, and A. Bezryadin, "Self-Assembled Wiggling Nano-Structures and the Principle of Maximum Entropy Production," ibid.5 (Feb 9 2015), https://doi.org/ 10.1038/srep08323).

Building upon these findings, Jeremy England is demonstrating dissipative adaption dynamics *in silico*. See J. M. Horowitz and J. L. England, "Spontaneous Fine-Tuning to Environment in Many-Species Chemical Reaction Networks," *PNAS* 114, no. 29 (Jul 18 2017), https://doi.org/10.1073/pnas.1700617114. Such research indicates pre-Darwinian (i.e., pre-biotic) adaptation at the purely material level. As adaptive, dissipative systems, such entities exhibit continuity with more complex evolving organisms, all of which utilize mutual information linking them to the energetic dynamics of their environment to generate and maintain themselves in the face of entropy. In a very real sense, then, such pre-biological entities engage in a learning process by which they acquire meaningful knowledge about their conditions that allows them to complexify. In a 2020 book, England writes:

> In essence, the idea of dissipative adaptation is that a system made of many pieces accesses energy from an external source differently depending on how those pieces are arranged, but that the energy that is absorbed also has an impact on how the arrangement of the pieces can change over time. Once that cycle of influence goes through multiple iterations, the opportunity arises for the shapes of things to skew strikingly in a direction that reflects the relationship between shape and energy flow in the system at hand. Sometimes, this means developing into a state that looks exceptionally good at grabbing energy, and sometimes it means the opposite: a robust state that is exceptionally skilled at avoiding the rearranging ravages of energy flow. (*Every Life Is on Fire: How Thermodynamics Explains the Origins of Living Things* (New York: Basic Books, 2020), 184).

In this sense, dissipative adaptation represents a form of "survival of the stable" much closer to that of classic biological adaptation.

When energy flows become more patterned, structures can emerge from those flows that make more efficient use of them for finding novel mechanisms of dissipation:

> We start with an inert lump of particles stuck in one shape, and we begin to subject it to a patterned driving of some kind—oscillation of a particular frequency, for example, or stimulation of some subset of chemical components. The energy flowing from the driving force powers an exploration of the space of possible configurations where the matter makes little local moves that change how its constituent parts are assembled relative to each other. If the interactions among the particles give rise to enough novelty through combinations of collective physical properties, then each small rearrangement can lead to changes in how the external drive energy gets absorbed, both in terms of how much energy flows and in terms of the kinds of motion the energy flow activates. As time passes, the system visits more and more states, and eventually it encounters an opportunity to settle into one that is exceptionally good at allowing the matter to stay in the same shape despite the constant physical insults it sustains from the external drive. This success at remaining stable in one particular shape or group of shapes is the result of energy absorption that is either sufficiently low, or powers motions that are exceptionally well matched to the assembled structure in ways that keep it from falling apart. Either way, by subjecting inanimate matter to patterned external drives, we ultimately expect to see the

> spontaneous emergence of a structure with energy flow properties that seem fine-tuned and special compared with those of a random rearrangement of the same building materials. (*Every Life Is on Fire*, 173-174)

[44] This view thus lies in stark contrast to conceptions of meaning as something evolutionarily late, ancillary, or limited to the symbolic plane of complexity. An example of such an account is Gary Tomlinson, *The Machines of Evolution and the Scope of Meaning* (New York: Zone Books, 2023). Tomlinson writes:

> [W]e come to understand ourselves, along with all other meaning-creating animals, as a swerve, late in evolutionary history, in the mostly meaningless processes governing life. This accident, described in this book as the irruption of processes that allow certain animals to create signs, doesn't set us apart from meaningless life-forms in any fundamental way. …Meaning is a small epiphenomenon in the earthly biosphere… (13)

Tomlinson explicitly situates his conception of meaning as post-structuralist, postmodern, and posthuman. Indeed, as the representative quote above signals, it is much more about *meaninglessness* than meaning, and profoundly misses the mark when it comes to positioning the semiotic meaning-making of humans within the broader context of semantic processing in nature.

On the other hand, the view of meaning embraced here would likewise avoid the other extreme: the one that places meaning and value as metaphysical primitives, somehow transcendently real and fundamentally preceding entity–field dynamics. Such conceptions, familiar from traditional religious and idealist philosophy, are sometimes courted in ostensibly post-postmodern projects to reconstruct meaning after postmodern deconstruction. For example, the 2024 work *First Principles and First Values* by Zak Stein, Marc Gafni, and Ken Wilber (published under the collective pseudonym David J. Temple) positions itself as being

> engaged in what we understand to be the great post-postmodern project, the reconstruction of value itself, honoring both those legitimate aspects of the critiques [of meaning and value] while also carefully and forcefully responding to the other, mistaken aspects. In doing so, we seek to restore the value that was always there—but not to its premodern condition. (David J. Temple, *First Principles and*

First Values: Forty-Two Propositions on Cosmoerotic Humanism, the Meta-Crisis, and the World to Come (Austin, TV: World Philosophy and Religion Press, 2024), 119–120)

To do so, however, they posit value as a "philosophical primitive" (165), axiomatically establishing it as a fundamental prerequisite to the universe itself. In their attempt to frame value in terms of "first principles," they declare that "Value is eternal in the sense that it is sourced beneath and beyond space and time" (199). Such an approach is corroborated, they say, by adopting an epistemological lens they call "Anthro-Ontology," whereby the human experience of value helps confirm its ontological basis. In all of this, their approach draws heavily from the tradition of "perennial philosophy," albeit one modified into an "evolving perennialism." As the authors point out, similar sorts of moves have also been made by philosopher Iain McGilchrist (see *The Matter with Things: Our Brains, Our Delusions, and the Unmaking of the World* (London: Perspectiva, 2023)).

While offering an important corrective to the postmodern approach, such efforts go, I believe, too far in the other direction, and ultimately fail to articulate a theory of meaning significantly different from its "premodern condition," which likewise simply posited meaning and value as transcendentally "sourced beneath and beyond space and time." A genuinely successful post-postmodern approach to meaning requires taking postmodernism's postmetaphysical turn seriously and so accounting for it *within* space and time. We do need a "first principles" approach to this issue; but rather than beg the question by simply positing value as such a principle, this book series turns to the first principles of the physical world—the laws of thermodynamics—as entirely sufficient for the emergence of meaning and value in the cosmos. Such an approach allows us to avoid the Scylla of meaning as a "small epiphenomenon" unique to sign-making creatures *and* the Charybdis of meaning and value as transcendentally sourced outside of the physical world of spacetime.

45 Azarian, *The Romance of Reality*, 38.

46 Terrence William Deacon, *Incomplete Nature: How Mind Emerged from Matter* (New York: W.W. Norton & Co., 2012), 23; italics removed.

47 Ibid., 25.

48 Ibid., 35. Naturalizing the teleological notion of "final causation" in biology is echoed by Henriques's observation about the need to counter reductive materialism with a so-called "*emergent naturalism*," since "dynamically self-organizing complex adaptive systems that engage in information processing and communication behaviors…require different

metaphysical concepts above and beyond efficient causation." Indeed, as suggested earlier, the rise of complexity science has offered a profound reconceptualization of Aristotle's four causes (material, efficient, formal, and final). Whereas earlier, reductionistic science met with considerable success by privileging material and efficient causation at the expense of the others, the impasse encountered when trying to extend this program into the sciences of life and mind eventually forced a paradigm shift. With the paradigm of complexity, formal cause returns through the lens of *constraint*, while final cause returns through the sorts of teleodynamics described by Deacon and others.

[49] Ibid.

[50] Indeed, this was the pioneering insight of physicist Erwin Schrödinger in his 1944 essay *What is Life?* that helped launch the far-from-equilibrium thermodynamic approach to adaptive systems and brought a thermodynamic perspective to biology. As he put it: "Thus the device by which an organism maintains itself stationary at a fairly high level of orderliness (= fairly low level of entropy) really consists in continually sucking orderliness from its environment" (Erwin Schrödinger, *What Is Life?* (Cambridge: Cambridge University Press, 2018 [1944]), 73). Such "negative entropy" (later shortened to "negentropy") allows order to increase locally even though disorder (entropy) always increases globally in compliance with the Second Law.

[51] The pioneering work of biologist Michael Levin is based on just these sorts of teleological/goal-based premises. See "The Computational Boundary of a "Self": Developmental Bioelectricity Drives Multicellularity and Scale-Free Cognition," *Frontiers in Psychology* 10 (2019), https://doi.org/10.3389/fpsyg.2019.02688, in which he offers a framing of cognitive entities as inherently goal-directed/teleological, writing:

> [W]hat defines a coherent, unified Self out of its constituent components and the surrounding environment is the set of parts that operate toward reaching specific goal states. Like a theorem, which has a holistic nature not shared by any of the axioms alone, a functional individual is more than the sum of its parts in the sense that goal-directed capacity emerges only from the integrated activity of all of the components. (Ibid., 7)

Selves, then, are emergent wholes maintaining a homeostasis of interdependent parts far from equilibrium who are driven to minimize anti-homeostatic stress.

This teleological drive fuels an ongoing and complexifying learning process. Levin writes:

> The initial, most primitive feature of a living system is preferences – the fact that some states of the world are better for its welfare than others. This enables learning from positive and negative reinforcement, which leads to an explosion of computational and behavioral capabilities. Preferences evolve into goals to the extent that a system grows in complexity and causal power and becomes able to act in the world in ways that are likely to move it toward preferred regions of its state space (initially, focused on simple metabolic survival, but ranging all the way to complex human psychological needs and perhaps beyond). (Ibid.)

As goal-directed systems complexify through learning, their "cognitive light cone" increases to encompass a broader landscape of past and future states informing its decision-making. "The cognitive boundary of an Individual," writes Levin,

> is the most distant (in time and space) set of events that this system can measure and attempts to regulate in its goal-directed activity. It is a surface indicating what things this system can possibly care about (conversely, it defines preferences as the spatio-temporal domain of states that serve as inputs and outputs to the system). In advanced agents, it is also the boundary of the self-model. (Ibid., 8)

As cognition complexifies, its horizons broaden in terms of what it can cognize and thus care about/attend to. That is, you could say, its scope of value expands, as its existence comes into contact and deepening relationship with more aspects of spatio-temporal reality.

The movement from tick to dog to human is thus one of expanding awareness and concern, as more of reality comes to impinge upon and inform its decision-making relative to achieving its goals. In short, the amount and kinds of meaningful information bearing on resisting entropy (i.e., minimizing anti-homeostatic stress) increases: meaning itself evolves.

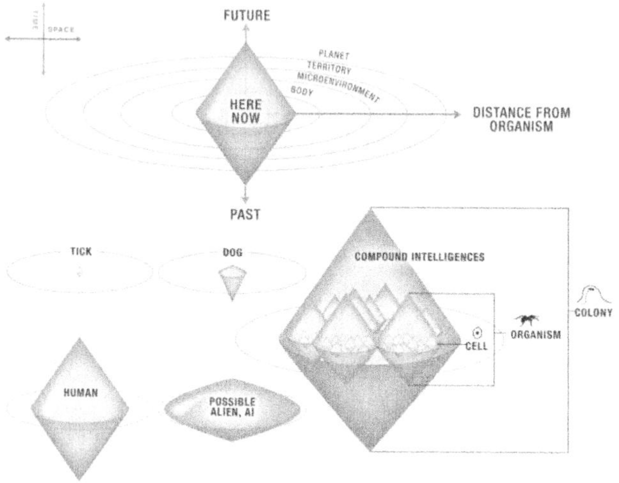

Source: Levin, "Computational Boundary," Figure 2.

[52] Terrence W. Deacon, "Shannon – Boltzmann – Darwin: Redefining Information (Part II)," *Cognitive semiotics* 2, no. 1 (2008): 180, https://doi.org/10.1515/cogsem.2008.2.spring2008.169.

[53] In his chapter "Natural Selection and Knowledge Creation" in *The Romance of Reality*, Azarian captures the full dynamics and import of Genetic Learning, writing:

> A statistical correlation between the configuration of the organism and the state of the environment means that the two systems share information. Building on pioneering work by the epistemologist-philosopher Fred Dretske, who applied Shannon's theory to natural selection a half century ago, Wolpert and Rovelli call this shared information *mutual information* (also known as "relative information"). As a species adapts to a niche, both in form and function, the amount of mutual information between the organism and its environment grows, such that we can say the adaptive system "learns" about its niche. For a species, growing more correlated with a niche means gaining more knowledge about that niche, which equates to becoming a better *predictor* of that niche. Semantic information is the mutual information that is causally necessary for the organism's survival. …[N]atural selection "transfers" information from the environment into the evolving biosphere,

> and…that information creates the intelligent, purposeful, goal-directed behavior that we associate with life and mind. (102-103)

Keep in mind, this occurs at the cellular level, and represents a form of "learning" in which genomes adapt to their environments through statistical correlation of information. Through the relentless variation and selective retention of successful organisms, nature prunes entropic noise from genetic information, creating organisms whose genetic information forms a hand-to-glove match with their environment. In a sense, evolution drives the encoding of genetic information the way a mold is made of an object. Just as the negative space of a mold reveals the object it copies, so Genetic Learning encodes a sort of map or model of an entity's environment.

As E. O. Wilson poetically put it, organisms "encode the predictable occurrences of nature's storms in the letters of their genes" (*The Diversity of Life* (Cambridge: Belknap Press, 1992), 9). That is, to use Azarian's formulation, reducing Shannon (informational) entropy, organisms reduce Boltzmann (configurational) entropy. The result is an increasingly refined internal model of an organism's niche encoded in the genome. As David Deutsch says, "genes embody knowledge about their niches" (*The Fabric of Reality: The Science of Parallel Universes and Its Implications* (New York: Penguin, 1997), 179). Or, as Azarian mnemonically summarizes it all: "*Adaptation = Statistical Correlation = Mutual Information = Model Optimization = Knowledge Creation*" (103).

It must wait for Cognitive Learning, however, before these internal models can become mental models in the way we tend to think of the idea. At the level of Genetic Learning, representing the world to the self is a process that rests latent in the genes; at the level of Cognitive Learning, representation becomes experiential, as a genuine simulation of the environment is produced through an integration of sensory inputs that allow the entity to navigate towards goal-states in real time.

[54] As Campbell summarizes what I am calling Genetic Learning in *The Knowing Universe*:

> Life's knowledge repository evolves through making new or abductive hypotheses, testing these hypotheses for their ability to produce viable phenotypes and retaining the genetic knowledge passing the test. The genetic variations expressed in a population's phenotypes are variously fit for existence and may come to exist in each succeeding generation at a different or updated frequency. In

each generation, the biological genomic model is updated and evolves to become more capable of existence.

This inferential view is consistent with our definition of life on earth:

> *Life is an inferential system in which a genetic model containing algorithmic autopoietic knowledge, accumulated by natural selection, exercises direct causal control over chemistry to create and regulate existing phenotypes.* (366)

[55] Deacon, "Shannon – Boltzmann – Darwin: Redefining Information (Part II)," 183.

[56] In a very real way, DNA encodes knowledge about the world in phenotypical expression. "In a completely literal sense," writes Azarian, "a dolphin's streamlined design, which is a product of the information stored in its genome, contains a knowledge of hydrodynamics. Similarly, an eagle's wing design contains a knowledge of aerodynamics" (*The Romance of Reality*, 96).

[57] Complexity theorist Stuart Kauffman reflects on this fundamental entwining of being and value, writing:

> A rudimentary beginning of "emotion" emerges here… The bacterium must sense its world and act to avoid toxins and to obtain food. The evaluation of "good" versus "bad," arguably the "first sense," enters here. …I note also that Hume's famous: "one cannot deduce an 'ought' from an 'is'", the famous naturalistic fallacy, rests on the critical fallacy that Hume, like Descartes, thought of a mind "knowing" its world. Hume did not think of an agent "acting" in its world. Given "action" and "doing", doing "it" well or poorly enters inevitably, and with it, "ought." With ought, the need for evaluation, the rudiments of emotion without positing consciousness, enter. ("Evolution Beyond Newton, Darwin, and Entailing Law: The Origin of Complexity in the Evolving Biosphere," in *Complexity and the Arrow of Time*, ed. Charles H. Lineweaver, Paul Davies, and Michael Ruse (Cambridge: Cambridge University Press, 2013), 169)

[58] Deacon, *Incomplete Nature*, 321, 322.

[59] Philosopher and theologian Donald A. Crosby captures this insight acutely in his 2023 book *Evolutionary Emergence of Purposive Goals and Values: A Naturalistic Teleology*, where he writes:

> "[V]alues do not so much already reside in nature as come into being within the *relations* of organisms to their environments. The locus of these values, in other words, is neither solely in nature itself nor in the creatures of nature but in the ways these creatures come to adjust and accommodate themselves to their environments and to one another in order to survive and thrive, and members of each species doing so in their own distinctive ways. Viewed in this way, there are no values as such in nature prior to the evolution of its multitudinous life forms. The values come into being as these life forms evolve and later in the evolutionary saga become able *consciously* to search for and find ways of adapting successfully to their environmental settings and to other life forms within those settings. (*Evolutionary Emergence of Purposive Goals and Values: A Naturalistic Teleology* (Albany: SUNY Press, 2023), 54; italics original)

While Crosby is entirely correct about the originary dynamics of meaning and value, he misses the fact that these dynamics shade down into pre-biotic levels, leading him to conclude that they emerge only with the emergence of life—rather than posit (as I do) that the naturalistic teleology of living organisms is itself an extension of more fundamental processes rooted in energy and information.

[60] Indeed, for the species to learn, the genome must be updated through selection pressures. Trial-and-error learning in this context means some organisms dying (failure in the problem space) and some living (success in the problem space). Krakauer reflects on this sobering reality, writing: "Darwin's theory of natural selection is a theory that introduces order into an individual genome by disordering a population. …The amount of information that is acquired for performing a preferred function is directly proportional to the loss of life" (David Krakauer, "The Inferential Evolution of Biological Complexity: Forgetting Nature by Learning to Nurture," in *Complexity and the Arrow of Time*, ed. Charles H. Lineweaver, Paul Davies, and Michael Ruse (Cambridge: Cambridge University Press, 2013), 232). In short, Genetic Learning is a phylogenetic learning process driven by the struggle for survival.

[61] If, according to the UTOK Tree of Knowledge, the modern theory of biological evolution (i.e., the evolutionary synthesis—or, I might suggest, the *extended* evolutionary synthesis) explains the processes at the "joint point" where Life branched off from Matter as a novel behavioral plane of complexity, then a comparable theoretical framework is needed to explain the processes that distinguish Mind from the Life plane. For this joint point, Henriques offers his own Behavioral Investment Theory

(BIT) to explain the phenomena we observe in the behavior of minded animals.

Fundamentally, BIT frames animal behavior in terms of thermodynamic cost-accounting. Henriques calls the first principle of BIT the *principle of energy economics*, "which is the notion that animals must, on the whole, acquire more workable energy from their behavioral investments than those behaviors cost" (Henriques, *A New Unified Theory of Psychology*, 48). Unlike plants or single-celled organisms, animals have complex nervous systems that coordinate their actions and movements within their environment and process the sensory information through the filter of energetic reward. All other aspects of animal behavior and the five other principles of BIT ultimately relate back to the logic of this energy economics.

The characteristics novel to the Mind plane are in many ways latent in the conceptions and language we use to describe the animal world. The word "animal" comes from the Latin *anima*, the *animating* principle believed to activate the body. Unlike rocks and even plants (relatively speaking), animals *move*, shuttling to and fro in pursuit of their various ends (e.g., food, mates, etc.). Behind such *movement* is *motivation*, the meaning behind the act. Animals have goals and are teleologically driven; they make energetic expenditures towards these goals in the form of behavioral investments. What separates them from organisms on only the Life plane of complexity is the nervous system, by means of which animals interface with their environment and learn how best to achieve their goals. Complex animals with subjective feeling *move* towards goal states based on *motivations*, which engender *emotions* relative to the success or failure of those goals. Henriques represents this through the formulation Perception – Motivation → Emotion (P – M → E). That is, "*perception of an actual state relative to a motivational state leads to an emotional state*" (Henriques, *A New Synthesis*, 338; italics original). In short, the thermodynamic imperative to stay far from equilibrium using energy from the environment, seen at the very base of the complexity stack, undergoes a series of transformations and developments from Matter to Life and into Mind, where the goal-directed movements of animal behavior, including expressions of emotion (happy, sad, angry, etc.) in response to goal states being achieved or thwarted, can be interpreted in light of its logic. In the last analysis, it is a logic of meaning, whereby purpose finds expression at varying degrees of evolutionary complexity.

[62] Krakauer observes: "A cognitive mechanism lends itself to exactly the same complexity definition that we used for evolutionary dynamics. ...Rather than a set of alternative genomes, consider a set of alternative

behaviors… The rate of update is no longer measured in units of generations but units of temporal action" ("The Inferential Evolution of Biological Complexity: Forgetting Nature by Learning to Nurture," 240).

[63] Richard Dawkins, *The Selfish Gene* (New York: Oxford University Press, 1976), 76.

[64] See Karl Friston, "The Free-Energy Principle: A Unified Brain Theory?," *Nat Rev Neurosci* 11, no. 2 (Feb 2010), https://doi.org/10.1038/nrn2787, https://www.ncbi.nlm.nih.gov/pubmed/20068583.

[65] For a useful introduction, see Thomas Parr, Giovanni Pezzulo, and Karl Friston, *Active Inference: The Free Energy Principle in Mind, Brain, and Behavior* (Cambridge, Massachusetts: The MIT Press, 2022). They summarize the notion of Active Inference as a goal-directed learning process that continually updates an entity's world model in response to how well it predicts (infers) future states. Accurate models allow organisms to predict with high probability the state of their environment, whereas inaccurate models will meet with surprise when expectations don't match reality, occasioning an update to the model. Through this process, the entity learns more and more about its surroundings. They write:

> Active Inference is a normative framework to characterize Bayes-optimal behavior and cognition in living organisms. Its normative character is evinced in the idea that all facets of behavior and cognition in living organisms follow a unique imperative: *minimizing the surprise of their sensory observations*. *Surprise* is to be interpreted in a technical sense: it measures how much an agent's current sensory observations differ from its preferred sensory observations—that is, those that preserve its integrity (e.g., for a fish, being in the water). Importantly, minimizing surprise is not something that can be done by passively observing the environment: rather, agents must adaptively *control* their action-perception loops to solicit desired sensory observations. This is the active bit of Active Inference. (6)

Or, more simply: "Surprise minimization permits living organisms to (temporarily) resist the second law of thermodynamics" (47).

As a theory of animal cognition, Active Inference extends into the domain of Mind the same sort of dynamics we have been tracking across the lower levels of cosmic complexification. The basic entity–field relationship creates a thermodynamic imperative from the start, setting off a cascade of increasing free energy dissipation, structural organization, and information generation characterizing a universal learning process that

grows more sophisticated from Matter to Life to Mind and beyond. At every level, there is meaning owing to the intrinsic desire of entities to extend their existence. So, too, at the level of Cognitive Learning, "Active Inference...is purposive and teleological" (57), a "normative," goal-driven process relating entities and environments in a field of value.

[66] See, for example, Francisco J. Varela, Evan Thompson, and Eleanor Rosch, *The Embodied Mind: Cognitive Science and Human Experience*, revised ed. (Cambridge: MIT Press, 2016), who helped frame the 4E approach to cognitive science by emphasizing the inherently inseparable and mutually-defining nature of organism and environment. Rather than understanding cognition as an entity's subjective mirroring of an objective world, we must see that mind and world shape one another. They write:

> Even the most hard-nosed biologist, however, would have to admit that there are many ways that the world is – indeed even many different worlds of experience – depending on the structure of the being involved and the kinds of distinctions it is able to make. And even if we restrict our attention to human cognition, there are many various ways in which the world can be taken to be. This nonobjectivist (and at best also nonsubjectivist) conviction is slowly growing in the study of cognition. As yet, however, this alternative orientation does not have a well-established name, for it is more of an umbrella that covers a relatively small group of people working in diverse fields. We propose as a name the term *enactive* to emphasize the growing conviction that cognition is not the representation of a pregiven world by a pregiven mind but is rather the enactment of a world and a mind on the basis of a history of the variety of actions that a being in the world performs. The *enactive approach* takes seriously, then, the philosophical critique of the idea that the mind is a mirror of nature but goes further by addressing this issue from within the heartland of science. (9)

Minds are, moreover, *embodied* in evolutionary entities—products of a developing and complexifying cosmos—who are *embedded* in the environmental contexts that shaped those minds, developing senses and perceptual acuities tailored to their particular niche and needs, meaning the world perceived by the mind is one *enacted* by the mind as much as representing it.

Such an approach offers, they suggest, a third way between the false binary of the "chicken position" ("the world out there has pregiven properties") and the "egg position" ("the cognitive system projects its

own world"). As they say, "Our intention is to bypass entirely this logical geography of inner versus outer by studying cognition not as recovery or projection but as embodied action" (172). Recalling the "figure-ground" distinction analogous to the bifurcation of entity and environment discussed in endnote 6, we might say that this approach appreciates the entire Gestalt of figure *and* ground as mutually defining aspects of one reciprocally defined whole.

Such a perspective leads to a profound appreciation for the variability of phenomenological consciousnesses rooted in different embodied, embedded, and enactive contexts. As they put it, "cognition depends upon the kinds of experience that come from having a body with various sensorimotor capacities, and…these individual sensorimotor capacities are themselves embedded in a more encompassing biological, psychological, and cultural context" (173). In short, entities operating at the level of organism–ecology, animal–environment, and person–society experience reality in profoundly different ways by virtue of being the sort of beings-in-context that they are. As a consequence, meaning will shift depending on these contexts, and upon the specific sensorimotor capacities of the entity in question. In their words, "the meaning of this or that interaction for a living system is not prescribed from outside but is the result of the organization and history of the system itself" (157).

[67] John Vervaeke, "Relevance Realization and the Emerging Framework in Cognitive Science," *Journal of Logic and Computation* 22, no. 1 (2012): 82, https://doi.org/10.1093/logcom/exp067.

[68] Kolchinsky and Wolpert, "Semantic Information," 9.

[69] Vervaeke, "Relevance Realization and the Emerging Framework in Cognitive Science," 86. Work by Andersen, Miller, and Vervaeke in 2022 shows how relevance realization can be mapped quite well to "precision-weighting" in the predictive processing model of Friston and others, wherein the level of surprise corresponds to how animals home in on relevant information and thus *meaning*. See Brett P. Andersen, Mark Miller, and John Vervaeke, "Predictive Processing and Relevance Realization: Exploring Convergent Solutions to the Frame Problem," *Phenomenology and the Cognitive Sciences* (2022), https://doi.org/10.1007/s11097-022-09850-6. Cf. Julian Kiverstein, Michael D. Kirchhoff, and Tom Froese, "The Problem of Meaning: The Free Energy Principle and Artificial Agency," *Frontiers in Neurorobotics* 16 (2022), https://doi.org/10.3389/fnbot.2022.844773.

[70] Such tradeoffs include things like balancing the tension between exploring novel affordances vs. exploiting given ones, focusing on certain motivations vs. diversifying aims, specializing in certain problem-solving

skillsets vs. becoming a more general problem solver, etc. See John Vervaeke and Leonardo Ferraro, "Relevance Realization and the Neuro-dynamics and Neuroconnectivity of General Intelligence," in *SmartData: Privacy Meets Evolutionary Robotics*, eds. Inman Harvey et al. (New York: Springer, 2013), 57–68.

[71] Fundamentally, they suggest, these opponent processes all come down to the same basic oppositional tension: namely, between maximizing *efficiency* vs. maximizing *resiliency*. Vervaeke and Leonardo Ferraro note that such dynamics correspond to actual wiring patterns in the brain. Rather than maximally resilient neuronal networks ("regular networks") or maximally efficient ones ("random networks"), optimal learning is accomplished through neural networks with aspects of *both* ("small world networks")—as has been shown to be a predictor of general intelligence (see N. Langer et al., "Functional Brain Network Efficiency Predicts Intelligence," *Human Brain Mapping* 33, no. 6 (Jun 2012), https://doi.org/10.1002/hbm.21297).

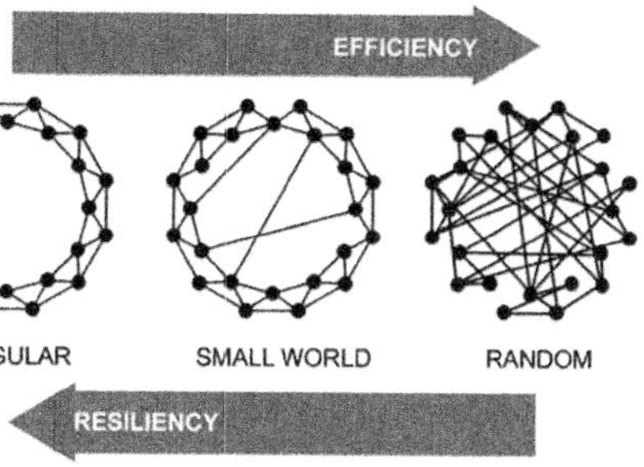

Source: Vervaeke and Ferraro, "Relevance Realization and the Neurodynamics and Neuroconnectivity of General Intelligence," Figure 6.6.

This balance between ordered regularity and chaotic randomness, you may note, is precisely what characterizes informational *complexity*. Meaning, it turns out, is best adjudicated at a certain critical threshold between maximal order and maximal chaos—the self-organizing criticality at the "edge of chaos" that characterizes *complex* systems.

[72] Vervaeke and Ferraro, "Relevance Realization," 63

[73] **Source:** Vervaeke and Ferraro, "Relevance Realization, Figure 6.3

[74] Vervaeke, "Relevance Realization and the Emerging Framework in Cognitive Science," 95.

[75] Simona Ginsburg and Eva Jablonka, *The Evolution of the Sensitive Soul: Learning and the Origins of Consciousness* (Cambridge: The MIT Press, 2019), 353-354.

[76] Ibid., 374.

[77] This integration represents a continuation of autopoietic, self-organizing and self-generating dynamics that operate on all scales of complexity. Such recursive, self-maintaining and self-preserving processes are, after all, the basis of teleological/goal-directed behavior. The constitution of an informationally distinct "individual" relative to environment is the basis of a self/not-self differentiation that produces meaning in the cosmos and sets the universal learning process in motion. We saw, for instance, how even at the level of Matter autocatalysis creates a closed loop whose causal chain engenders a self-referential basis for normativity. At the level of Life, these dynamics continue, but operate now at the more complex level of the cell. The cell becomes the informational individual serving as the self-reference point for normativity. When cells become networked into a nervous system for multi-cellular organisms, this reference point shifts again. The new whole is now the animal of the Mind plane, which is made up of multiple sensorimotor mechanisms whose normative concern becomes the integrated totality of the animal. It is at this level that conscious experience enters the picture, because consciousness is essentially the self-reference point of the animal to itself within the animal's own world model.

Azarian writes: "For consciousness to arise…a system must not only model its environment, it must also *model itself*" (*The Romance of Reality*, 203). As Douglas Hofstadter put it, the "I" is a "strange loop," a recursive representing of the representer to itself. In his chapter on "Consciousness" in *Incomplete Nature*, Deacon speaks of "a self that creates within itself a teleodynamic reproduction of itself" (528). One way of thinking about this is as a higher-order self-reference point for the system as a whole, by means of which it can better reach its teleological ends of staying far from equilibrium. In preconscious cognition, or purely "associative learning," the animal whole is more of a collection of relatively autonomous subsystems who dynamically reinforce the whole through synergistic activity even while remaining meaningfully distinct. Consciousness emerges when these subsystems become truly unified under the organization of a higher-order self-reference point. The integration of such differentiation is a complexification process that creates a complex system with emergent properties: namely, consciousness.

Like earlier forms of meaning-driven and hence normative entities, such consciousness is inherently value-based; that is what valence qualia *are*. Deacon concludes:

> So why is there a "what it feels like" associated with this neural teleodynamics? And why does this "being here" have an intrinsic good and bad feel about it? The answer is that this self-similarity-maintaining dynamic provides a constant invariant reference with respect to which all other dynamical regularities and disturbances are organized. Like the teleodynamics of autogenic self, it is what organizes all dynamics around an invariant *telos*: the self-creating constraints that make the work of this self-creation possible. (*Incomplete Nature*, 530)

This integration of an organism's normative neural subsystems into a higher-order self-system becomes the new self-reference point directing the entity's behavior. As minded animals pursue their goals relative to a mental model they are continually updating through experience, this higher-order self enters the world model as the organizing referent for all behavior. In this way, consciousness emerges as "self-awareness," at least in the sense that the conscious animal experiences positive and negative qualia as an integrated whole. That whole is the sentient "I," the first-person perspective. Consciousness is inherently normative.

[78] The sorts of minded animals that occupy the complexity plane Mind, then, can be divided into unconscious and conscious categories. Henriques refers to the cognitive apparatus related to the former as Mind^1 and the latter as Mind^2. Purely reflex-based reaction cognition describes the workings of Mind^1. This kind of cognition can occur without any subjective consciousness, the way your body is currently pumping blood through your arteries and regulating your breathing and other vital functions without any conscious awareness or intention on your part. The nervous system simply takes in information, processes it, and generates a suitable output behavior in response in line with its teleological, homeostatic goal-states far from equilibrium. Even without any conscious awareness, let alone linguistic-conceptual knowledge, Mind^1 identifies the information that is relevant for the functioning of the animal and hones its reflexive responses in a *meaningful* way. As Henriques summarizes it:

> One way to conceive of cognition…in functional terms is that it is the semantic meaning the neurocognitive system generates to frame the animal–environment relation and identify what is relevant in the

> landscape of affordances that might guide its path of behavioral investment. (*A New Synthesis*, 291-292)

Mind^2, by contrast, exhibits subjective conscious experience (SCE), wherein those semantic meanings are integrated and coordinated. Surveying a broad literature on the evolution and nature of consciousness, Henriques summarizes:

> SCE is likely not present in the reacting, reflexive, fixed action level of animal mentation, such as seen in a sea slug. The argument here is that the beginnings of SCE emerge with the learning level, such that pleasure and pain can be thought of as nature's first qualia that integrate and coordinate flexible learning patterns based on association and consequence. These 'mental flashes' model the animal–environment relationship and orient approach or avoidance behaviors. (*A New Synthesis*, 373-374)

[79] On the inherently emotive and thus normative basis of consciousness, see also the work of Antonio Damasio, such as *The Feeling of What Happens: Body and Emotion in the Making of Consciousness* (New York: Harcourt Brace, 1999) and *The Strange Order of Things: Life, Feeling, and the Making of the Cultures* (New York: Pantheon Books, 2018).

[80] This is an important distinction, because the proponents of IIT have framed the theory as an explanation of consciousness by simply conflating integrated information with phenomenological awareness. The *central identity* claim of IIT, for instance, declares "Any experience is identical to the maximally irreducible cause-effect structure associated with the system in that state." But correlation and causation, let alone identity, should not be confused. Such philosophical overreach has unfortunately made IIT vulnerable to critique, even as "pseudoscience"; see I. Fleming et al., "The Integrated Information Theory of Consciousness as Pseudoscience," (2023), https://doi.org/10.31234/osf.io/zsr78.

[81] A strong positive correlation between intelligence (IQ) and neuronal complexity has been demonstrated by N. A. Goriounova et al., "Large and Fast Human Pyramidal Neurons Associate with Intelligence," *eLife* 7 (2018), https://doi.org/10.7554/eLife.41714.

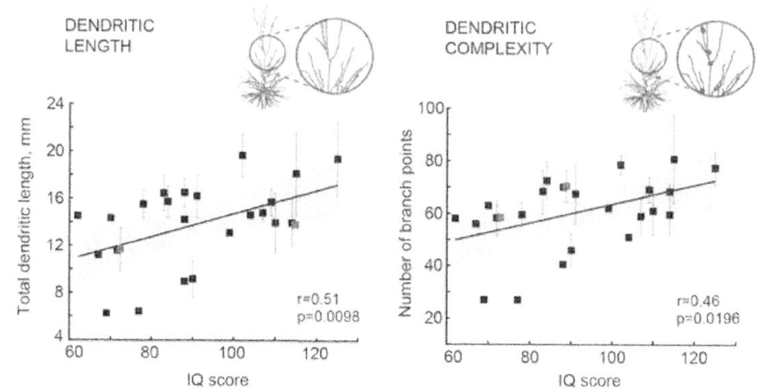

Source: Goriounova et al. 2018, Figure 3; adapted.

Wedded to findings that show correlation between IQ and the thickness of cortex tissue (see, e.g., K. L. Narr et al., "Relationships between IQ and Regional Cortical Gray Matter Thickness in Healthy Adults," *Cereb Cortex* 17, no. 9 (Sep 2007), https://doi.org/10.1093/cercor/bhl125), this suggests that more neuronal complexity directly relates to general intelligence.

Evidence that more neuronal complexity also relates to greater degrees of consciousness include studies that map integrated information to states of awareness (awake, asleep, anaesthetized, etc.). For instance, by sending a magnetic pulse cascading through the brain, researchers can assess the complexity of its path using the sort of algorithmic complexity metric discussed in endnote 19. This "zap-and-zip" method offers a way to measure the complexity of the brain's neural network in informational terms. The method is summarized in Christof Koch, *The Feeling of Life Itself: Why Consciousness Is Widespread but Can't Be Computed* (Cambridge, MA: MIT Press, 2019), where Koch explains:

> Think of cortex as a large brass bell and the magnetic coil as the clapper. Once struck, a well-cast bell will ring at its characteristic pitch for a considerable time. And so does the awake cortex, buzzing at a variety of frequencies. In contrast, the brain of someone in a deep sleep acts like a stunted or cracked bell… Although the neurons remain active, as evidenced by the strong, local response, integration has broken down. …Ultimately, each person's EEG response is mapped onto a single number, the pertubational complexity index (PCI). If the brain barely reacts to the magnetic pulse—say, because the EEG is nearly flat (as in deep

> coma), its compressibility will be high [i.e., little irreducible complexity] and PCI will be close to zero. Conversely, the maximal complexity yields a PCI of one. The larger the PCI, the more diverse the brain's response to the magnetic pulse, the more difficult it is to compress [i.e., the more complex it is].

This method was introduced by A. G. Casali et al., "A Theoretically Based Index of Consciousness Independent of Sensory Processing and Behavior," *Sci Transl Med* 5, no. 198 (Aug 14 2013), https://doi.org/10.1126/scitranslmed.3006294. Summarizing their findings, they write:

> We test PCI on a large data set…recorded in healthy subjects during wakefulness, dreaming, nonrapid eye movement sleep, and different levels of sedation induced by anesthetic agents (midazolam, xenon, and propofol), as well as in patients who had emerged from coma (vegetative state, minimally conscious state, and locked-in syndrome). PCI reliably discriminated the level of consciousness in single individuals during wakefulness, sleep, and anesthesia, as well as in patients who had emerged from coma and recovered a minimal level of consciousness. (1)

In short, increases in integrated information in the brain (measured using algorithmic complexity) correspond to increases in conscious awareness:

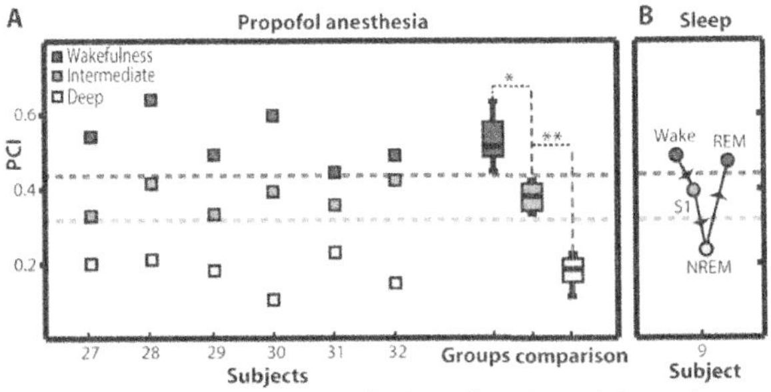

Source: Casali et al., "A Theoretically Based Index of Consciousness," Figure 4.

In sum, more intelligence = more neuronal complexity = more consciousness. As their nervous systems complexify, minded animals become more conscious and better learners. Such complexification is driven by evolution, as "evolution selects organisms with high Φ^{max}"

(Koch, *The Feeling of Life Itself*, 124). *In silico* experiments have produced evidence of this, which show that as environmental complexity increases, the integration of information in adaptive entities increases. Thus, the learning process itself naturally drives the development and expansion of consciousness. See L. Albantakis et al., "Evolution of Integrated Causal Structures in Animats Exposed to Environments of Increasing Complexity," *PLoS Comput Biol* 10, no. 12 (Dec 2014), https://doi.org/10.1371/journal.pcbi.1003966.

[82] **Source:** Koch, *The Feeling of Life Itself*, Figure 11.2; adapted.

[83] The example of the meaningfulness of a frisbee to a dog gives us some opening insight into the sorts of ways meaning complexifies. Simple functionalist interpretations tend to reduce meaning to biological necessity. To say cheese is meaningful to a mouse can be understood more readily in such functionalist terms. The mouse needs the cheese to survive, so it is meaningful to it. The survival of a dog does not directly depend upon frisbees, however. The meaningful connection is more indirect and complex. Rather, the survival of a dog is dependent upon their relationship with their human owner, such that the dog's survival is implicated in social dynamics in a way the mouse's is not. Dog owners enjoy playing frisbee with their dogs, the dog's ability to play frisbee is an exaptation of other skills it developed through its species' evolution, and the net outcome is a productive symbiotic relationship for both dog and human. This complex network of interactions is what is meaningful to the dog, reified in the form of the frisbee. In this case, the dog has meanings it cannot fully cognize, since it cannot trace the full network of relationships that impinge on its viability. This complexity only becomes amplified at the level of human culture, where the degree of indirect meaning is greatly increased. Indeed, human enculturation and development is largely about individuals learning to appreciate more of the webs of relation in which they participate. The viability of the human individual is dependent upon the viability of the social collective. Today, as human causal power has grown, and human actions have direct causal impact on the biosphere, human learning and development means becoming aware of the ways in which their viability is directly related to the viability of the entire global ecosystem. This represents the cutting edge of adaptive pressure on *Homo sapiens* in the 21st century, and the development of knowledge frameworks that can facilitate this learning process is the driving edge of cultural evolution (see the discussion of metamodernism in a forthcoming volume).

[84] Thomas Nagel, "What Is It Like to Be a Bat?," *The Philosophical Review* 83, no. 4 (1974): 439.

[85] Actually, one could trace these insights even further back still, to Ludwig Wittgenstein, who recognized, for similar reasons: "If a lion could talk, we could not understand him."

[86] Von Uexküll is the scientist most cited by Martin Heidegger, for instance, and also influenced Merleau-Ponty, Foucault, Deleuze and Guattari. On the cybernetics front, his idea of "function-cycles" was a forerunner of the idea of "feedback loops" in systems theory, and both Ludwig von Bertalanffy and Humberto Maturana draw from his work.

[87] Jakob von Uexküll, *A Foray into the Worlds of Animals and Humans: With a Theory of Meaning*, trans. Joseph D. O'Neil (Minneapolis: University of Minnesota Press, 2010 [1934]), 144.

[88] Again, cf. Levin, "The Computational Boundary of a "Self"."

[89] As David A. Crosby puts it:

> The bee values its queen, hive, nectar, and blossoms. The lion values its hunt and its cubs; the beaver its aspen trees, dams, and den; the bacterium its sugar; the tree its sun, soil, and rain; all enough to devote their lives—consciously or unconsciously—to these necessarily valued goals and ideals. Values in nature are neither problematic, rare, or absent. Instead they are omnipresent in all forms of life. There would be no facts about life without a pervasive immanence of the pursuit of values in all living creatures. (*Evolutionary Emergence of Purposive Goals and Values*, 61)

[90] Ludwig Feuerbach, *The Essence of Christianity*, trans. George Eliot (New York: Cosimo Classics, 2008 [1902]), 8.

[91] There is evidence, for instance, that such advanced mammals "mourn" or "grieve" the death of their kin. See Sanjeeta Sharma Pokharel, Nachiketha Sharma, and Raman Sukumar, "Viewing the Rare through Public Lenses: Insights into Dead Calf Carrying and Other Thanatological Responses in Asian Elephants Using Youtube Videos," *Royal Society Open Science* 9, no. 5 (2022), https://doi.org/10.1098/rsos.211740; G. Bearzi et al., "Whale and Dolphin Behavioural Responses to Dead Conspecifics," *Zoology (Jena)* 128 (Jun 2018), https://doi.org/10.1016/j.zool.2018.05.003.

[92] **Source:** C. B. Stringer, "Evolution of Early Humans," in *The Cambridge Encyclopedia of Human Evolution*, ed. Steve Jones, R. D. Martin, and David R. Pilbeam (Cambridge: Cambridge University Press, 1994), 242.

[93] Henriques, *A New Synthesis*, 246.

[94] Terrence W. Deacon, *The Symbolic Species: The Co-Evolution of Language and the Brain* (New York: W.W. Norton, 1997), 340.

[95] Specifically, the teacher-student relationship, most likely originating in kin communities (e.g., parent-child teaching contexts). See Kevin N. Laland, *Darwin's Unfinished Symphony: How Culture Made the Human Mind* (Princeton: Princeton University Press, 2017), especially the chapter "Why We Alone Have Language."

[96] Another, related reason is that human children are not fully physically developed at birth like other animals. The incredibly large head of *Homo sapiens* resulting from runaway encephalization cannot pass through the birth canal fully formed, so all humans are born "premature" and continue to grow to cerebral maturity outside the womb. Consequently, infants must be reared longer, which in turn supports enculturation and development of socio-linguistic learning capacities.

[97] The period when language moved from ostensive communication with symbols to robust grammatical constructions of them is difficult to discern. If *Homo habilis* did have some form of language, for instance, it was very likely not a full grammatical one. Did the phase transition to grammatical propositional language begin c. 300,000 years ago, or as late as the "cognitive revolution" of c. 70,000 to 30,000 BCE? Was the transition slow or sudden? These details, though important and fascinating, need not concern us here.

[98] Henriques, *A New Synthesis*, 87-88.

[99] Henriques, *A New Unified Theory of Psychology*, 119.

[100] Henriques, *A New Synthesis*, 50; italics original.

[101] Such self-conscious awareness became increasingly selected for owing to the social pressure to *justify* behavior among other linguistically-endowed beings. Since all animal social relationships are freighted with meaningful information important to survival of the individual, gaining influence and fostering supportive relationships among the group is a normative aspiration for all social creatures. In human contexts, these social dynamics get overlaid with a linguistic layer. Self-conscious linguistic justification thus becomes a key mechanism by which human social agents learn to achieve their own normative goals within the context of the group.

This insight is absolutely vital, for it shows why human meaning cannot be understood as myopically self-serving despite the fact that meaning, fundamentally, is about entities preserving themselves in an entropic world. Faced with such an account, some might balk at the prospect that meaning is ultimately about self-preservation, since often the human ideas and beliefs held to be most meaningful (e.g., love, generosity, self-sacrifice, etc.) are precisely *not* self-focused but emphatically concerned with the well-being of *others*.

These concerns are obviated when we appreciate that what makes us human is precisely our enculturation through and into human community. We become persons by participation in the transpersonal categories of language. The concerns of the group are *our* concerns; our well-being hinges on its well-being; collective meaning and individual meaning are inherently interwoven. While this is true to some degree even for social animals, the nature of symbolic language radicalizes the co-constituting relationship of self and society through the open information channel of inter-subjective communication via justification.

[102] Such a "voice in one's head," one will note, sounds a lot like the "angel on one's shoulder," which is to say: one's *conscience.* As we shall explore in greater detail in the next book, the ethical impulse (like all meaning and value) is not bestowed transcendently from without but rather emerges transjectively from within the immanent world—in this case, through the accommodation to the meanings of other agents. The moral imperative emerges in human Culture as linguistically-mediated egos form through the internalization of inter-subjective values. On the Culture plane of complexity, to be an "I" requires a social "We" through which I get enculturated. In this way, your meanings and values shape my own, and vice versa. Through language, I cognize the subjectivity of the Other—a process that can then be extended to the more-than-human world. Crosby writes:

> Our moral obligations to [human and] nonhuman creatures have their basis in the natural fact of our necessary relations with one another. There is no need for further or deeper explanation of their binding power. We are committed to the presence of other-values because we must be. It is true that we can endeavor to overlook or be oblivious to this fact, but we do so only at the cost of our own endangerment as a species of life… (*Evolutionary Emergence of Purposive Goals and Values*, 66)

Contrary to certain theological framings, which see morality as inexplicable within an evolutionary framework, we are led to conclude that our own viability is dependent upon others. All meaning is ultimately shared meaning; all value is ultimately interconnected in the web of complexity.

[103] **Source:** Gregg Henriques, "The Three Filters of Consciousness," *Psychology Today*, https://www.psychologytoday.com/us/blog/theory-knowledge/202203/the-three-filters-consciousness.

[104] That inner voice demanding justifications of the self-narrating ego represents an internalization of the social world within my very psyche. The ego itself is inextricably interwoven with the collective. The "I" is inherently a "We."

[105] Urie Bronfenbrenner, *The Ecology of Human Development: Experiments by Nature and Design* (Cambridge: Harvard University Press, 1979).

[106] **Source:** "Ecological-Theory-1024×1024-1," © Deirdre Budzyna via. ROTEL, licensed under a CC BY-SA (Attribution ShareAlike) license.

[107] Azarian, *The Romance of Reality*, 175.

[108] Ibid., 176.

[109] Enculturation is thus a transmission process of symbolic information on the person–society plane of complexity. Like other learning processes, it is subject to a continual filtration process that can "update" justification systems as environmental conditions change. Transmission always entail variation, and with variation comes context-dependent selection.

[110] Azarian draws on Karl Popper's "evolutionary epistemology" to frame science specifically in this light. In science, multiple hypotheses are developed and tested against the evidence, with only those that effectively predicted the outcome being selectively retained. Such dynamics match those of evolution's variation and selective retention algorithm, itself a learning process. While agreeing with Azarian, we must emphasize that science is only one (admittedly succinct and efficient) form of this process, which also characterizes the development of prescientific worldviews. Indeed, modern science represents a method of learning that itself had to be learned through such a process. For this reason, we must see all of human cultural evolution as a learning process.

[111] Rovelli, "Meaning and Intentionality = Information + Evolution," 18.

[112] Ibid., 24.

[113] Campbell frames his articulation of a "knowing Universe" in terms of a universalization of the "genotype–phenotype" relationship characteristic of Genetic Learning through the free energy principle (à la Karl Friston). He sees this playing out on the quantum-to-classical level, the biological level, the neuronal-computational level, and the cultural level (very much mirroring independently the UTOK Matter, Life, Mind, and Culture planes of information processing). At the quantum level of this learning processes, information transfer leading to selective retention occurs from the decoherence/collapse of the quantum wave function from quantum probabilities to classical actualities. Campbell summarizes:

> [I]t is the network of exchanged and recorded quantum information that composes classical reality. In this view, classical reality emerges from quantum information recorded in each quantum system's wave function. ...[W]e may consider the quantum wave function as a control mechanism on the unfolding of physical reality in the same sense that we can consider the genome as a control mechanism on the developmental unfolding of biological phenotypes. (*The Knowing Universe*, 255, 257)

This notion of universal Darwinism is still speculative, so I have parenthetically placed the idea of "wave function" information processing as a hypothetical basis of the universal learning process.

[114] The parallels between such a chart and the famous "hierarchy of human needs" by Abraham Maslow should not escape comment. Existing at the top of the complexity stack, humans have material, biological, animal, and cultural needs based on the meaningful goal states of our evolutionary history through Matter, Life, Mind, and Culture. Maslow's hierarchy captures this insight, with physiological and safety needs forming the basis of needs for belonging, self-esteem, and "self-actualization."

[115] 1 Corinthians 15:26.

[116] Wallace Stevens, "Sunday Morning," line 63.

[117] Hence the "paradigmatic" nature of the sacred as emphasized in the comparative religious studies of Mircea Eliade. According to Eliade, man as *Homo religiosus* relates to the sacred as the perfect template, the ideal type, the primordial reality, and through religious rites returns to that mode of things as they were "*ab origine, in illo tempore.*" The sacred is the *archetypal* source of strength and power—an enduring repository, continually open for renewing engagement, for enhancing viability and vitality:

> The man of the archaic societies tends to live as much as possible *in* the sacred or in close proximity to consecrated objects. The tendency is perfectly understandable, because, for primitives as for the man of all pre-modern societies, the *sacred* is equivalent to a power, and, in the last analysis, to *reality*. The sacred is saturated with *being*. Sacred power means reality and at the same time enduringness and efficacity. ...Thus it is easy to understand that religious man deeply desires *to be*, to participate in *reality*, to be saturated with

> power. (Mircea Eliade, *The Sacred and the Profane: The Nature of Religion* (New York: Harcourt, 1959), 12-13)

The sacred is the deep source of paradigmatic reality received through intergenerational transmission. Conforming behavior to its template maximizes *being* and *power* (i.e., viability).

[118] The idea of the sacred as shaping a cultural "world model" finds deep consonance with the theoretical literature, I think. A religious world is a *cosmos*, a framework for reality that gives meaning and orientation to life. This is exemplified in Eliade's writings, for instance, who notes:

> [W]e could say that the experience of sacred space makes possible the "founding of the world": where the sacred manifests itself in space, *the real unveils itself*, the world comes into existence. ...[T]the world becomes apprehensible as world, as cosmos, in the measure in which it reveals itself as a sacred world. ...Religious man thirsts for *being*. His terror of the chaos that surrounds his inhabited world corresponds to his terror of nothingness. The unknown space that extends beyond his world—an uncosmicized because unconsecrated space, a mere amorphous extent into which no orientation has yet been projected, and hence in which no structure has yet arisen—for religious man, this profane space represents absolute nonbeing. If, by some evil chance, he strays into it, he feels emptied of his ontic substance, as if he were dissolving in Chaos, and he finally dies. (ibid., 63-64.)

The world-constituting sacred provides the blueprint for the symbolic, linguistically-mediated, transpersonally realized cosmos in which meaningful human existence unfolds. Beyond that order is entropic chaos—a framing we will return to in great depth in following books.

For further consideration of the sacred as the basis of "world formation," see William E. Paden, *Religious Worlds: The Comparative Study of Religion*, 2nd ed. (Boston: Beacon Press, 1994) As the organizing framework for his comparative analysis, Paden writes: "Religions create, maintain, and oppose worlds. Their mythic symbols declare what the world is based on, what its oppositional forces are, what hidden worlds lie beyond or within ordinary life. ...It is of the very essence of religion to interpret the world" (53-54). Simply put, "the sacred is the foundation of a world" (61). The sacred world carves out an intelligible sphere for human activity that enshrines moral knowledge and collective values as teleological lodestars for guiding the behavior of its members. The

individual "downloads" the worldview of their culture as part of ego development, whereby the *telos* of the individual and the collective become shared.

[119] This insight allows us to appreciate the truths contained in Émile Durkheim's seminal sociological reading of the sacred—namely, as the enshrinement of the transpersonal force facilitating social cohesion and generative integration of the collective. In worshipping the shared sacred object of reverence (be it totem, spirit, god, what have you), humans are actually renewing society itself: the superorganism (we might say) sustaining and preserving every individual as parts within the whole.

In his classic 1912 work *The Elementary Forms of Religious Life*, Durkheim posits:

> We can say, in fact, that the worshipper is not deluding himself when he believes in the existence of a higher power from which he derives his best self: that power exists, and it is society. When the Australian is transported beyond himself and feels life flowing in him with an intensity that surprises him, he is not prey to illusion. That exaltation is real, and it is really the product of forces external and superior to the individual. Of course he is mistaken when he believes that this heightened vitality is the work of a power that takes plant or animal form. But his error lies only in taking literally the symbol that represents this being to men's minds, or the form of its existence. Behind these figures and metaphors, crude or refined, there is a concrete and living reality.
>
> Religion takes on a meaning and a logic that the most intransigent rationalist cannot fail to recognize. The main purpose of religion is not to provide a representation of the natural world… [R]eligion is above all a system of notions by which individuals imagine the society to which they belong and their obscure yet intimate relations with that society. This is its primordial role; and although this representation is metaphorical and symbolic, it is not inaccurate. Quite the contrary, it fully expresses the most essential aspect of the relations between the individual and society. For it is an eternal truth that something exists outside of us that is greater than we are, and with which we commune.
>
> That is why we can be sure that acts of worship, whatever they might be, are not futile or meaningless gestures. By seeming to strengthen the ties between the worshipper and his god, they really strengthen the ties that bind the individual to his society, since god is merely the symbolic expression of society. (Émile Durkheim, *The*

Elementary Forms of Religious Life (Oxford: Oxford University Press, 2001), 170-171)

In our terms, we could say that the sacred represents the shared meanings of the emergent system of Culture, which unite human thought and action towards viability-enhancing goals. Though Durkheim goes too far in presenting what amounts to a reductionist account of the sacred ("god is *merely*..."), he does capture a profound truth, and one we shall return to in subsequent books.

[120] Eliade relates a story from the anthropological account of the Australian Aborigines by Spencer and Gillen which has continued to haunt my memory. He writes:

> According to the traditions of an Arunta tribe, the Achilpa, in mythical times the divine being Numbakula cosmicized their future territory, created their Ancestor, and established their institutions. From the trunk of a gum tree Numbakula fashioned the sacred pole (*kauwa-auwa*) and, after anointing it with blood, climbed it and disappeared into the sky. This pole represents the cosmic axis, for it is around the sacred pole that territory becomes habitable, hence is transformed into a world. ...Spencer and Gillen report that once, when the pole was broken, the entire clan were in consternation; they wandered about aimlessly for a time, and finally lay down on the ground together and waited for death to overtake them. (*The Sacred and the Profane: The Nature of Religion*, 32-33)

Today, we would call such an experience a "meaning crisis."

[121] We can understand the insights of comparative mythologists like Joseph Campbell and Jordan Peterson precisely in this light, who offered theories of a "monomyth" and "metamythology," respectively. In these theories, the meaning of every hero story is actually about meaning itself: exploring the unknown, battling chaos, and thereby renewing the world (i.e., updating the world model). For an insightful piece connecting complexification, John Vervaeke's relevance realization, and Jordan Peterson's "order vs. chaos" framework from *Maps of Meaning*, see Brett P. Andersen, "Intimations of a New Worldview," *Substack*, September 4, 2022, https://brettandersen.substack.com/p/intimations. We will be exploring these sorts of connections further in following books.

[122] Another thing this perspective affords is the ability to see "the sacred" in continuity with all forms of meaning to have emerged across cosmic complexification. Perhaps it is best to keep the idea of the

"sacred" as a purely Cultural term, with referents only to linguistically-mediated symbolic information. But it is not difficult to see how complex animals would have their own experiences of awe and transcendence as natural products of learning dynamics. Some interpret instances of animal behavior (e.g., bears sitting and gazing from vistas, monkeys swinging through the air over waterfalls) as evidence of wonder in other mammals. When we appreciate that such positive phenomenological states are the product of cognitive minds attempting to accommodate to an overwhelm of information, this seems eminently plausible.

123 With the advent of language and Symbolic Learning, the capacities for humans to update their world model exponentially increased. Indeed, it now became possible to entertain the idea of *ultimate meaning* in the sense of acquiring a truly comprehensive and all-encompassing vision of reality. Such a grand vision would be sacred indeed, as it would allow one optimal insight into the nature of the reality we all must navigate to maximize viability.

A helpful way to think about this is through increasingly expanding scales of contextualization. If it is valuable for me to accurately map myself in relation to my immediate environment (e.g., modeling the room I am in to better navigate it), there will also be utility in mapping my room in relation to a still wider environment (e.g., knowing how the room is situated within the larger contexts of my immediate property, my town, my nation, etc.). Such information helps contextualize my immediate agent–arena dynamic in a way that informs how I navigate within it, and how I would navigate out of and beyond it. This is a purely spatial way to frame the issue, but in fact we must consider it as metaphorical for all the increasingly abstract and comprehensive framings we might have of reality "as a whole" (cf. Levin's "cognitive light cones" and the extension of the horizons of one's normative goals; endnote 51).

The more such a map expands away from my immediate situation, the less immediately relevant it is to me, true—but that does not make it negligible or superfluous information. Quantum mechanics is an incredibly abstract and distant terrain compared to knowing where my lunch is. And yet, the behavior of my iPhone renders it literally at hand and pragmatically impinging, while my own ego identity has had to be constructed by means of and in relation to my culture's dominant world model, such that 'I' am co-defined and enacted in relation to concepts about subatomic particles and superpositions. As noted above, the inherently interconnected nature of a complex universe means that information across all levels interweave and influence my existence (and my survival) in ways that would give me greater causal power if I could

more accurately conceptualize them. Knowing, then, how "it all fits together" or "what it's all about" to contextualize all the contexts I can imagine would indeed be of great utility and *value.*

It is, I believe, the valued goal-state of attaining such a big picture perspective that informs our normative quest for not just meaning *in* life (which is always contextually given by our local transjective relationship dynamics) but a sense of meaning *to* life (which places all such local meanings in a broader global frame and thus serves as a meta-normative guide). Such, I believe, is the function of so-called "meta-narratives," which make meaning of our meanings by placing all knowledge within a grand teleological frame.

As we shall see, however, no metanarrative is ever actually complete. All represent moments in the collective human learning process. As such, each retains its own conception of the sacred, which both coheres its gains and drives its expansion into the unknown. The sacred of one metanarrative is the stepping stone to the sacred of a still more advanced symbolic world. Through this developmental process, meaning and value are forever being transcended.

[124] Feuerbach, *The Essence of Christianity*, 17. Such a thought calls to mind the famous saying of the pre-Socratic philosopher Xenophanes, who quipped that if horses could make art, they would make *their* gods in the form of horses. But if Xenophanes was critiquing humans' tendency to anthropomorphize divinity, Feuerbach was celebrating it, feeling new vistas opened up by the prospect of relating to the human psyche as it revealed itself in religion and theology. Both point to a fundamental psychological insight we will be exploring in greater depth over the next two volumes: the fundamentally assimilatory nature of learning minds.

[125] Ibid., 38-39.

[126] No doubt such language rings with a certain unavoidably Hegelian resonance. Hegel was the first great thinker to consider human spirituality through a developmental lens. Without Hegel, Feuerbach would never have uttered the words above, and without Feuerbach, Marx would not have turned Hegel's dialectical idealism into dialectical materialism. There is clearly a rich lineage of such modes of thought, and certainly this work would not exist without *all* of the minds just mentioned. That said, in getting our bearings better, it should be stated plainly that this work accords perfectly comfortably with none of these thinkers. In some senses, it has elements of all of them.

[127] See the next volume in this series, *Psyche and Symbolic Learning* (Sky Meadow Press, forthcoming).

142

Printed in Dunstable, United Kingdom

80391635R00088